Oaklore

Jules Acton

Illustrations by Sally Mollan

Oaklore

Adventures in a World of Extraordinary Trees

GREYSTONE BOOKS
Vancouver/Berkeley/London

Text copyright © 2024 by Jules Acton
Illustrations copyright © 2024 by Sally Mollan

24 25 26 27 28   5 4 3 2 1

All rights reserved. No part of this book may be reproduced, stored in a retrieval system or transmitted, in any form or by any means, without the prior written consent of the publisher or a licence from The Canadian Copyright Licensing Agency (Access Copyright). For a copyright licence, visit accesscopyright.ca or call toll free to 1-800-893-5777.

Greystone Books Ltd.
greystonebooks.com

Cataloguing data available from Library and Archives Canada
ISBN 978-1-77164-966-7 (cloth)
ISBN 978-1-77164-967-4 (epub)

Developmental editing by Nancy Flight
Editing by Lesley Cameron
Copy editing by Elizabeth Peters
Proofreading by Robert Sharman
Indexing by Stephen Ullstrom
Jacket design and composite by
Jessica Sullivan and Javana Boothe
Jacket images (birds): The Natural History Museum
and Florilegius/Alamy Stock Photo
Text design by Jessica Sullivan
Interior illustrations by Sally Mollan. Page 21 based on photograph courtesy of Andy Lindsley; page 36 based on photograph courtesy of Graham Calow; page 160 based on photograph courtesy of Christoph Benisch.

On page 132, quote from Peter Wohlleben, *The Hidden Life of Trees*. English translation copyright © 2016 by Jane Billinghurst. Reprinted by permission of Greystone Books.

Printed and bound in the UK on FSC® certified paper at CPI Group Ltd. The FSC® label means that materials used for the product have been responsibly sourced.

Greystone Books thanks the Canada Council for the Arts, the British Columbia Arts Council, the Province of British Columbia through the Book Publishing Tax Credit, and the Government of Canada for supporting our publishing activities.

All URLs were live at the time of writing.

Greystone Books gratefully acknowledges the x̱ʷməθkʷəy̓əm (Musqueam), Sḵwx̱wú7mesh (Squamish), and səlilwətaɬ (Tsleil-Waututh) peoples on whose land our Vancouver head office is located.

For Jeanie and Susie, for your great love of the world and for your kindness to its occupants

Contents

Introduction 1

··· 1 ···
DOORWAYS TO OTHER WORLDS 13

··· 2 ···
WASPS, WORDS AND OAKSPIRATIONS 36

··· 3 ···
CROWNING GLORIES 61

··· 4 ···
LIFE, DEATH AND BEETLING ABOUT 87

··· 5 ···
ENCHANTED FORESTS: FOLKLORE,
MYTH AND MAGIC 109

··· 6 ···
INCREDIBLE EDIBLES: HEALTH AND HEALING 132

··· 7 ···
THREATS: COULD A BEAUTY BE A BEAST? 160

··· 8 ···
THE OAK'S LITTLE HELPERS.
AND ITS BIG HELPERS. 185

Conclusion 211
Some Fun Stuff 213
Acknowledgements 227
Endnotes 229
Select Bibliography 245
Index 247

Introduction

HERE IN THE UK WE ARE BOTH TREE POOR AND TREE RICH. ON THE ONE HAND ONLY 13% OF THE UK IS COVERED BY TREES. THAT IS SHOCKING AND WE NEED TO CHANGE IT. YET WE HAVE GREAT TREASURE TOO IN OUR OLDEST TREES, WITH MORE REALLY BIG, ANCIENT OAKS THAN THE REST OF EUROPE. AND TO KNOW THESE LIVING LEGENDS IS TO LOVE THEM.

DR DARREN MOORCROFT, CHIEF EXECUTIVE, WOODLAND TRUST, 2024

EXTRAORDINARY ORDINARY THINGS

People across the UK love our oak trees – and why wouldn't we? They are easy trees to love; they are all around us, in our woods, in our fields and hedgerows, alongside our roads and footpaths. They are among our most common trees and perhaps those our children find easiest to identify, thanks to their wavy, wavy leaves and their iconic, air-polished acorns. Our history and our culture have evolved with oak trees; we have met under oaks, preached under oaks, and made magic with oaks. We have hidden special things in our oak trees: everything from the spoils from a hunt to royalty to – in legend, at least – Robin Hood. We have created great buildings from oaks. We have fought wars with ships made from oaks and, back in 1620, a few of us set off to find the 'new world' of North America in our oaken vessels. We instinctively understand that if you want to associate your organisation with strength, resilience, permanence and culture, you could do worse than to choose an oak for your logo; the forest of oak-themed logos is a story in itself. The roots of our oak trees and the human activities they support spread wide indeed.

And then there is the sheer, beguiling beauty of oak trees. A glance at any oak – even a young one standing modestly at the side of a street – is enough to lift the soul a notch. And, as the oak matures, the experience gets bigger and richer; to glance up through great, muscular, oaken limbs at a pale winter's sky is to experience majesty.

Stay with an old oak a little, sit alongside it, and you will, with a spot of luck, see something move in its branches, a creature flitting between leaves, perhaps, or a teeny, tiny dot of a thing – a shiny little beetle, maybe – peeking out from a crevice in the bark, suddenly to disappear back into the folds of the tree's trunk. Really look into the oak tree and a glorious thing happens, a new world opens up, a universe of animals and plants and fungi and bacteria, all of them getting on with the important stuff of life: looking for a home, raising their young, sniffing out something to eat, and also, on occasion, finding something with which to pick a bit of a fight. Because an older oak is not only a tree, it is a nature reserve all by itself. Oaks in the UK provide food, shelter, housing and support for an astonishing 2,300 other plants, animals and fungi, who, according to Dr Ruth Mitchell and a team of biodiversity experts at the James Hutton Institute, 'use oaks as living space or feed in them, or roost in them, or breed in them'.[1]

And these are just the species we have counted.

The species in question include birds, spiders and beetles – lots and lots of beetles, some of them sparkling like jewels, others dark like the night. There are butterflies and stiletto flies and moths and mosses and flapworts and featherworts.

There are fungi, then more fungi, really loads and loads of fungi, that strange class of things that are neither plant nor animal, but arguably more animal than plant (weird, I know – jump straight to Chapter 6 if you need to know more immediately). There are fungi that love living wood, fungi that love dead wood, fungi that love dry things, fungi that love wet things. There are fungi that most of us have never heard of and there are – almost certainly – many fungi that haven't yet been discovered. Then there are things that are arguably even stranger than fungi, like lichens, lots of lovely lichens. There is nothing like a lichen for weird wonderfulness.

If we break down the 2,300 species that use the oak by numbers, we're counting at least 38 bird species, 31 mammals, 108 fungi, 1,178 invertebrates, 716 lichens and 229 bryophytes (mosses, liverworts and hornworts). Of these, 320 species depend upon oak trees for their very existence; these beings simply couldn't, well, be, without them. For the others the oak is a key part of their world, their habitat or their diet.

And, while one tree alone probably won't house all 2,300 of these characters – that is a collective task – the amount of activity going on in each tree is astonishing. A mature oak is a community, and it is a community with which we human beings are intimately connected.

GETTING GRIPPED BY LIFE IN OAKS

Now then, the thing about some of the characters in this oak world is that they aren't always easy to meet; as entomologist Professor Brian Eversham says, 'The wildlife of trees tends not to come buzzing out at you in the way that of a meadow would.'[2] In a project that started during the Covid pandemic, during the periods between lockdowns I set out to find some of them; to get to know their stories a little and to try to understand the way their worlds meet ours. This became a kind of nature-and-culture ramble, which, on the invitation of the lovely Rob Sanders of Greystone Books, ended up in this book.

I needed help to access the world of the oak, a lot of help from people like Ruth Mitchell and Brian Eversham and many others. They became guides; wise men, women and children who helped bring into view the mysterious lives of the little things in the big, big oak. With their support I also discovered activities that would help me get up close and personal with oaks and the species that depend upon them, that revolve around them and that link together into communities. I have shared these activities in this book so that you can enjoy them too if you choose.

When it comes to the species featured in the book I haven't tried to be comprehensive – to do that would probably involve living as long as the oldest oak and my comparative pipsqueak of a human life is too short even for the number of species we know the oak embraces, let alone those we don't yet know about. I have featured just a few. Some of the species highlighted are already well known and iconic; others are, I think, more obscure and surprising. Some are already closely linked with oak trees in our culture; others less so. Most of the characters and stories featured have made it into the book simply because I have met them along the way and have come to love them a little.

For more detail on oaks, and their place in nature and culture, I recommend the books of Archie Miles, Aljos Farjon, William Bryant Logan and Julian Hight. For a further celebration of everything oak I also applaud the books of James Canton and John Lewis-Stempel. There are other sources of inspiration listed in the bibliography – and, of course, from our wonderful conservation organisations whose incredible passionate staff and volunteers are driving the charge to save our oak trees and our nature.

'Save?' you might ask. 'Save them from what?' The sad truth is that oaks and their world face many threats. These range from disease to insensitive development. But, before we despair, we should also know there are many ways to help them. Thankfully, many of the ways of helping the oak are also fascinating and fun.

Helping nature tends to start with loving it. And, along the way, this quest to find out more about oaks and their world turned into a love story: the love story between oak trees, their world and we humans. It is a tale of community and the glorious, beautiful, fragile connectedness of all of the things that enable us to live and which go a long way towards making life worth living.

Now, as we all know, the course of true love never did run smooth, to quote William Shakespeare (who, like many writers, has a debt to oak trees that I'll cover in the book). I'm afraid I can't promise you an easy ride of a read at all times. There are threats, danger, and lots of jeopardy for the oak and its world, but if love is at the heart of our relationship with oak trees, it is also love that will save them, for save them we must.

In terms of geography, the focus of the book is the UK and its native species of oaks. There are also some references to Brits, Britishness and the British Isles. Often the term I've chosen to use is for accuracy, and reflects how it appears in the source material. This sometimes makes most sense in a historical context as political geography has changed over the centuries. National identities and the way we refer to them can of course be a sensitive thing, and I hope this approach doesn't cause any offence to anyone or cause anyone to feel excluded – that is the last of my intentions. What I'm aiming for is to embrace the oaks on the islands in this here neck of the woods. That said, much of the book was written at the peak of the Covid pandemic; I wasn't able to travel as widely as I'd have liked, so there is an English – even Midlands – slant to the narrative, but thankfully I managed to venture a bit further at times and met some stunning oaks on the way. What I have aimed to do here is simply share stories and vignettes, alongside ways of getting up close and personal with oak trees and the world around them. I hope they will inspire more people to get out and about to enjoy their own nearby oaks.

I'm an oak enthusiast rather than an oak expert. I have been lucky enough to have encountered many brilliant and learned people in my working life in conservation charities, and I find that if you hang around the experts, some of their knowledge will stick to you like pollen to the knee of a bee. And I'm delighted to be able to share here what I have learned from those experts.

I have also come to understand that the more you learn about nature, the more you realise how much you don't know. There are always more worlds to explore, more connections to make. That is – in part – the fun of it.

Wishing you all the joy of the oak and its worlds.

· · · Know Your Oaks · · ·

COMMON IN A GOOD WAY

The oak itself is relatively common. There are around 121 million of them in UK woodland.[3] There are also lots of oaks outside of woods, which means that a great many people, if we choose, can enjoy exploring one in a nearby park or at the side of a path. Almost one million oaks live in London alone.[4]

Two kinds of oak are native to the UK: the pedunculate oak and the sessile oak.[5] Each has an impressive array of aliases. The pedunculate oak is also called the English oak, the common oak, the European oak and, if you want to get scientific about it, *Quercus robur*. The sessile oak, meanwhile, can go by the name of the Irish oak, the Welsh oak, the Cornish oak, the durmast oak or *Quercus petraea* (*quercus* meaning 'oak'). I tend to use the terms pedunculate and sessile.

Whatever we call them, they are grand things, these plants. And we also have other, non-native oaks that people have brought in – relatives such as the red oak, the holm oak and the Turkey oak.⁶

Two native oaks on a relatively small landmass. You might think this makes the UK special. And it does, kind of. Although at this point we should also acknowledge that there are hundreds of other types of oak in the world. Literally hundreds. Some say there are around 400 species, others more – I have read claims for up to 600.⁷ Oak trees grace the lives of people across the Northern Hemisphere, from Canada to Colombia to China.

As Woodland Trust chief executive Darren Moorcroft points out, there is something stand-out special about UK oaks, and that is the sheer number of them that fall into the category of 'ancient'. An astonishing number of our oak trees are categorised as ancient, at 400 years old or more.⁸ Some could be 900 years old, and a rare few are even thought to be 1,000 years old or more. While it is difficult to be precise about the age of old trees (chopping them down to count the rings would be both pointless because of the way old oaks hollow out and, of course, deeply wrong), we think some of these tree beasts could date back to William the Conqueror and even King Offa.

Girth is one indicator of age and, of the surviving really, really big old oak trees – those with trunks nine metres or more in circumference – we have recorded more than 100 in the UK, compared to 85 in the whole of mainland Europe.⁹ The UK's richness in ancient trees is all the more surprising given that we are, in general, a shockingly tree-poor set of islands; 13% woodland cover, as things currently stand, is pretty dismal compared to more than 30% in most European countries.¹⁰

OAK SPOTTING

Even from a distance, mature oaks have a distinctive appearance. They have an appealing bunchy, roundness. As TV presenter Simon King says, an oak is 'like a cloud, a big, bulbous spongy cloud; that's because the oak has very fine twigs and leaflets spreading in all directions at the end of every branch and it produces this wonderful, rounded outline unlike almost any other tree in Britain'.[11] He's not kidding when he says they are big – mature oaks can grow to 30 metres (100 feet) tall or more (although when they get really tall they tend to shrink back down in height; more on that later in the book).

How to tell our native oaks apart on the horizon? Each has a slightly different shape; the sessile oak is usually narrower than the pedunculate, so people who really know their trees can often distinguish them in this way.[12] However, that takes a practised eye.

An easier way to distinguish sessile oaks from pedunculate oaks is to get up close at autumn time and check out their acorns. There is a clue in the names (although, in all honesty, it isn't the most helpful one unless you happen to be a fan of classical languages). Peduncle means dangly bit or stalk. Sessile means sitting or seated. With your pedunculate oak, the acorns are attached to the twig by a long stalk – a dangly bit, we might say. If, as a child, you ever pretended acorn caps and their tiny twigs were little pipes, I suspect you'd have been playing with a pedunculate. The sessile oak is less dangly when it comes to the acorns. Its seedy bits have teeny, tiny short stalks, and also cluster together in pairs. The letter 'S' can help as a bit of a memory aid though, if you think of 'Sitting Sessile'. Sessile oaks are often found more to the north and west of the UK and in uplands, but not always.

I have made up an acorn-based ditty that might help:

An acorn that's sitting
is an acorn most fitting
a sessile's north westerly style.
But acorns that dangle,
like charms from a bangle,
suggest a peduncle meanwhile.

Easy, right? Bob's your peduncle.

Leaves can also help you tell which oak is which. However – confusion alert – here the whole dangly system gets flipped. Your pedunculate oak leaves, it turns out, are not danglers. Not at all. They have almost no stalk. Helpfully, though, they do have a distinctive bulge that looks like an earlobe, at the base of the leaf near the stalk. In contrast your sessile is a better dangler in the leaf department, with a long stalk. Its leaves are a bit more jaggedy around the edges.

Got that?

Just when you think might be getting the hang of all this, you find out that our oaks hybridise – cross-breed – so mixing up all the patterns we have just got into our heads. These hybrids are so common they have their own scientific name, *Quercus x rosacea* (for the uninitiated the 'x' there makes it clear it is a hybrid).

In short, it isn't *always* easy to tell your pedunculate from your sessile, nor either of your native oaks from your quirky hybrids. But, if this is beginning to noodle your noddle, don't worry, we shouldn't get too hung up on this. Oaks are all glorious and they all support an amazing web of life.

ANCIENT, VETERAN AND NOTABLE OAKS

Given some luck, actually a lot of luck, an oak tree will live up to the old saying that an oak grows for 300 years, lives for 300 years and dies for 300 years.

We'll explore the reasons for the relative richness in ancient treasure later in this book but why does it matter? Why are these oak trees so important? Well, it's not only because of their age and the history they have lived through – these trees are living monuments – but, also that their longevity makes for a relatively stable environment over time in and around which other species can survive and thrive.

The Ancient Tree Inventory (ATI) is a great online source of information mapping the oldest and most important trees in the UK. On it trees are classified as ancient, veteran or notable and, together, they total more than 190,000, many of which are oaks.[13] These are just the trees we have found. We know there are many others out there. Many, many more.

Ancient trees are trees that are old *for their species* – it varies considerably. In the case of the long-living oak, as we have seen, ancient oaks are those that are 400 years old or more. Ancient oaks are typically found in hedgerows, parkland, designed landscapes, fields, woodland and wood pasture. Occasionally we come across them in avenues, streets and gardens.

Many ancient oaks are greater than six metres in girth and can get as big as ten metres, or even a little more.

Characteristics of ancient oaks, with thanks to the Ancient Tree Inventory for information, include:
- a large girth (circumference);
- major trunk cavities or progressive hollowing;
- decay holes;
- physical damage to trunk;
- bark loss;
- large quantities of dead wood in the canopy;
- crevices in the bark, under branches or on the root plate, sheltered from direct rainfall;

- fungal fruiting bodies (from heart-rotting fungi species – more on these later);
- a high number of interdependent wildlife species.

Ancient oak trees may also have:
- a pollard form (pollarding is the regular cutting of upper branches) or physical indications of past management;
- a prominent position in the landscape;
- cultural or historic value.

A veteran tree, according to the ATI, is a 'survivor that has developed some of the features found on an ancient tree, not necessarily as a consequence of time, but of its life or environment'.

We also talk about 'notable trees'. These are usually mature trees that stand out in their local environment because they are large by comparison with other trees around them.

The websites of the Ancient Tree Inventory and the Ancient Tree Forum[14] have great tips for assessing the age of a tree. I recommend having a look, not least to find out more about the really special trees near you. They might also inspire you to go on an ancient tree hunt of your own.

1
Doorways to Other Worlds

OURS IS AN ISLE OF VOICES WHOSE MILD AIR
AND GENTLE SKIES ARE SWEETENED EVERYWHERE
WITH A WINGED MUSIC...

FRANCIS BRETT YOUNG, FROM *THE ISLAND*

BEGINNING WITH THE BIRDS

Oak trees, so folklore tells us, are doorways to other worlds. Other worlds held quite an appeal during a Covid-infested winter while our own world was coughing, spluttering and grinding its way to a halt.

So it was that my dog friend, Pepe, and I found ourselves in the heart of Sherwood Forest on a misty February morning when the world was in turmoil. The day was warming up. The year was waking up. I was embarking on a quest to discover more about the mystery and magic of oaks and the doors they can open up for us.

A great tit chimed like a tiny bell, high up in an oak branch. Blue tits lisped along in tinselly harmony and a gang of long-tailed tits, which sounded like they were unzipping anoraks, added a light percussion. They buzzed to the beat of the night-fallen rain, which now dripped heavily, steadily from the tips of the trees.

Out of the blue, a drum roll. A great spotted woodpecker burst onto the sound scene with its battering blast, the woodland's answer to Phil Collins.[1] I sought out the feathery musician, peering up into misty treetops through muscular tree limbs that reached into the milky haze and tapered on and up and out into fine, fingery twigs.

Suddenly we had a glimpse of the drummer bird, uniformed as it was in smart black, white and red. For its instrument it had chosen something perfectly percussive: a thick, dead oaken branch that clawed, antler-like, from the tree.

The sight and sound of this creature gave me a little bird-fuelled lift. It made the journey worthwhile; more than worthwhile, in fact: a delight. I had wanted to see some nearby woodpeckers because of the special role they play in nature, but we were in one of those pandemic periods when we weren't allowed to meet in groups, so I couldn't just take myself down to the nearby visitor centre to ask where I should look. Instead, I phoned the Royal Society for the Protection of Birds (RSPB), which manages the site, and got hold of Carl Cornish, conservation officer. Carl patiently explained the way over the phone: 'go past The Bee Tree, up to The Major Oak, head out past The Medusa Oak...' All instructions should sound like this, I thought, as if you are a character in *The Lord of the Rings* and Gandalf is sending you off to a new, strange world, one of which you have heard fantastical tales, to find something precious. In a great many ways, that was indeed the case.

This being February, woodpeckers seemed good companions with whom to begin an adventure exploring the life in and around oak trees. I had heard their drumming is intense around January to April. And, fittingly, like oak trees, woodpeckers are homemakers, providing accommodation to all sorts of other creatures by filling trees with little holes, cavities in which those other creatures set up house.

Of the 2,300 species that are supported by oak trees in one way or another, many, I had learned, enjoy a woodpecker nook. They include other birds, bees, wasps and the red-listed (i.e. threatened) Bechstein's bat. Later I was to go on up to peep through some of these creatures' doorways and find out more about their lives, but first I felt compelled to linger a while to get to know woodpecker world a little, mainly because I'd fallen a little in love with woodpeckers. Woodpeckers, it turns out, are amazing.

Woodpeckers are musicians. They are carpenters. They have been called tree surgeons and guardians of the forest.[2] Their importance in nature is reflected in their status as a 'keystone species' and 'umbrella species'; essentially, they support other beings that, in turn, help them to exist. They have inspired stories, poetry and even the design of bicycle crash helmets. Actually, the latter might have been based on a misunderstanding of woodpecker physiology, but, still, they are beautiful, they are diverse; some are quite strange and, like all the best families, they have a cousin who comes with a whiff of witchcraft. And if all that wasn't enough, they have such extraordinary tongues that they once prompted Leonardo da Vinci, back in 1490, or thereabouts, to write a note on his to-do list: 'Describe the tongue of the woodpecker.'[3]

Unfortunately, da Vinci's description of said tongue seems to have been lost – if he ever got around to writing it – but thankfully others have studied these woodpecker lickers. Thanks to those dedicated people, we know that those tongues are remarkable for several reasons, not least of which is their astonishing length. The great spotted woodpecker ('or great spot'), for example, has an appendage that can pop about four centimetres (1 1/2 inches) out of its beak – great for grabbing a grub and impressive for a creature that is only about 22cm (nine inches) long at adulthood.[4] By my calculations, that means a human blessed with an organ of similar proportions would be able to stick it out about 28 centimetres – that's about a foot for those of you who like imperial measures. A foot of tongue hanging out of a human mouth. Let's be thankful evolution didn't take that route.

Conveniently, perhaps because of Mother and Father Nature's tendency to be organised, the woodpecker's tongue is retractable, a little like the electric cable on a vacuum cleaner, and, when not in use, it wraps itself neatly inside the woodpecker's head. This system is not only a tidy and convenient arrangement, it is also multifunctional: the tongue and its supporting structures act like a handy inner cushion. They appear to help protect the bird's noggin from scrambling like an egg as it hammers out its tune against a branch, at 10 or more strikes per second in the case of the great spot.[5] This might not be the only safety feature built into this incredible animal's head. Some have argued that spongy cartilage inside a woodpecker head acts as a shock absorber, and this is the theory that helped inspire the design of the aforesaid bicycle helmet.[6] Others counter it does no such thing and the pecker's brain is protected simply because of the size of the bird, but, still, they inspired the design of a crash helmet. Well done woodpeckers.[7]

Ever more amazed by the fantasticness of woodpeckers, I delved into Gerard Gorman's book *Woodpeckers* and online sites like that of the Woodpecker Network, and everything I read increased my sense of wonder.[8] So I hope you'll bear with me while we spend some time diving further into this particular woody world here. I learned there are three 'main' woodpeckers in the UK. They are our great spotted friend, *Dendrocopos major*, the lesser spotted woodpecker, *Dryobates minor* (also known as *Dendrocopos minor*), and the green woodpecker, *Picus viridis*.

The green woodpecker is notable for many things, not least for inspiring Professor Yaffle – whom some will remember from the loveable TV programme *Bagpuss*. It also wins my award for best nicknames. These include nicker pecker, yaffle bird, weather cock, yappingale, wet bird, Jack Eikle and, a personal favourite, laughing Betsey. It also sports a lovely outfit: a mossy green coat and stunning red cap, which makes it fairly easy to identify by sight. The two spotted versions are more easily mixed up with each other, or at least they would be if the little lesser spotted versions weren't so rare. They both bear smart black, white-chevroned livery, accessorised, in some cases, with dabs of red. To help tell them apart, the Woodland Trust's Amy Lewis points out that the lesser spotted has 'barring all the way across the back rather than the "shoulder" patches of the great spotted. They also lack the red area beneath the tail.' The lesser spotted male is much like the female but 'with a red crown'.[9]

In addition to these three, there is also their strange and exciting cousin, the wryneck, *Jynx torquilla*, who appears on the scene occasionally. Interestingly, the wryneck isn't on the list of 2,300 species mentioned above, but we know it isn't a stranger to the oak tree. It deserves a brief mention here for its visits to the UK – even if it generally doesn't stop to breed here these days, preferring other parts of Europe. It is hard to see the family resemblance between the wryneck and its cousins, and it isn't a drummer. Instead it has a different talent: an ability to writhe, hiss and turn its head almost 180 degrees. Gerard Gorman says, 'Currently the most common explanation for this bizarre behaviour is that it is mimicry, a direct imitation by the bird of the movements of a snake when threatened, but, tempting as this explanation is, no one has yet been able to offer firm evidence to prove it.'[10] In the past, this strange behaviour has led to superstitious people associating the bird with misfortune and magic.[11]

Personally I'm having no truck with the idea of bad luck. I would feel fortunate indeed to see a wryneck writhing away in our neck of the woods. Wrynecks used to breed in the UK, but they are no longer thought of as residents. On this, the British Trust for Ornithology (BTO) says, 'The decline and extinction of this species in the UK is believed to have been driven by a drop in food availability caused by a shortage of bare ground and short vegetation, although a number of other factors may have contributed including agricultural intensification, climate change, an increase in conifer plantations and the effects of pesticides.'[12]

The green and spotted woodpeckers, thankfully, can be found in Sherwood Forest and, on subsequent visits, as well as being drummed by the great spot, I have stood in a clearing as the yaffle sound of the green woodpecker whoops and bounces off the tree trunks. This can make you feel a little like you are being laughed at, which is fine by me. If I'm going to feel mocked I'd like it done by wildlife. I haven't yet had a personal encounter with the rare lesser spotted woodpecker in Sherwood – that would be a much cooler, much more impressive experience in birding terms – but Carl and his colleagues at the RSPB tell me they are there and so now it is on my list. Yes, I am now the proud keeper of a birdy bucket list, something I didn't even realise I needed until I started this quest. But that's the thing about new openings, new worlds: you find you want to go further and deeper. You find yourself making lists of wild things.

Despite the rarity of the lesser spot, woodpeckers on the whole are winners in evolutionary terms. Says Gorman, 'They have one of the most extraordinary anatomies in the animal world, every aspect of which is pitched towards working on wood',[13] and Charles Darwin remarked that the woodpecker's *'feet, tail, beak and tongue … are so admirably adapted to catch insects under the bark of trees'* [original italic].[14] That tongue again. Along with other woodpecker features, it has clearly caught the imagination of several great minds while playing its own role in the survival of the fittest.

As homebuilders, or rather home-excavators, the woodpeckers are hard workers, workers of wood. They can spend

20 days making a nest hole, chipping away like feathery little carpenters on the sunny side of a tree. And, adorably, they seem to have their own little dust masks for the job: tufty bits at the base of their bill appear to filter dust and splinters, helping keep their eyes clear.[15] After all that effort to make a home, you might think they'd move in, settle down happily and be done with it, but no, they tend to move each year, creating a new hole a bit higher up the tree and going through the whole carpentry routine anew. There is a reason for this, though. Like humans – well, most humans – they don't go through the upheaval of moving house just for fun. They move probably so they can avoid the build-up of pesky pests in their homes: fleas and lice and other things that make you feel itchy with a mere mention. So the hard work and upheaval all have a greater purpose.

Of course, nothing goes to waste in nature. More holes means more homes; more homes for the whole community, and this is one of the reasons woodpeckers have been called ecosystem engineers. As their fellow engineer Leonardo da Vinci is believed to have said, 'Learn how to see. Realise that everything connects to everything else.'

MORE FEATHERY OAK LOVERS

We have seen that UK oak trees support at least 38 species of birds by offering them a place to live, perch or eat, or all of the above. And – here's one way things connect up – some of our woodlands' star birds are grateful for old woodpecker nooks and crannies.

The pied flycatcher (*Ficedula hypoleuca*) is one of them. For me, getting a glimpse of this bird involved an excursion way beyond my home base of Nottinghamshire, to Bryn Arw, a heavenly hillside in the Brecon Beacons in south-east Wales. I was there with Dr Keith Powell, a seventh-generation Black Mountains farmer, and Robert Penn, a best-selling author. The two men are on a mission to plant native, broadleaf trees across Bryn Arw, with the help of the local community, in a project called Stump Up For Trees.[16]

Our pied flycatcher announced its presence with a sound that I once read described on an RSPB blog site gorgeously as 'like a bumpy squeaky train, bouncing along for a few seconds, then hitting the buffers'. Then it flew out from a stunted-looking, wind-battered sessile oak. Being a male, it was looking all dapper and dressed up, with most of its feathers a glossy black and white, giving off a two-tone vibe like a member of a 1980s ska band. Had it been female it would have been paler, less flashy. (This aesthetic difference between the sexes – flashy males, less-than-flashy females – not uncommon in the bird world, might at first seem a bit unfair on the females until you realise that blending into the background can be a good survival tactic, especially when you are sitting on your eggs.)

While oaks and housing holes are important to pied flycatchers, they are only part of their particular world, because these little creatures are intrepid international travellers. They fly all the way from West Africa to the UK and back each year. What hazards they must overcome en route! And what a feat of endurance for an animal that, at around 13cm (about five inches) in length, is about as long as a pair of spectacles is wide and which weighs about 13 grammes (about half an ounce). To put that in perspective, it is about the same weight as a standard AAA battery. They make the epic journey to our mature woods in late April, mainly in the west of the UK and particularly in Wales, to breed – which arguably makes the UK their home base – and head off to warmer climes in September.[17]

As you can imagine, these weary, weeny travellers need a really good feed at the end of their journey to the UK. They love to feast on the caterpillars of our oaky moths, so it is particularly important that there is a plentiful supply there in readiness for their return home in the spring.

Another bird partial to a disused woodpecker hole is the nuthatch (*Sitta europaea*). Carl tells me that, as they move in, nuthatches do a spot of DIY; they 'customise the entrance, using mud to make a front door that is their size'. They are the bird world's answer to Laurence Llewelyn-Bowen. Nuthatch style doesn't end there. Have you seen nuthatch eyes? They remind me of the song 'Jeepers Creepers' with its nods to wonderful 'peepers'. They have this streak that looks like a kind of elongated slash of gothic eyeliner, worthy of the super-glamorous singers Siouxsie Sioux or Lady Gaga. Theirs must be one of the chicest faces in the British bird business, making the nuthatch, in my view, a rock-and-roll kind of a bird. They can also blast out a top tune. Their riffs go along the lines of *chit-chit-chit*, a loud and rapid *twit-twit-twit-twit* or slower *sirr-sirr-sirr* and a gleeful, piping *whee, whee, whee*.[18] Another loveable thing about nuthatches is that they have their own funny walk that goes against the usual system: they often take a walk headfirst down a tree. This little habit might make the nuthatch look like it's doing an audition for a Monty Python sketch but it is actually an excellent tactic for spotting insects other birds haven't noticed. Nuthatches are great inspiration for people who like to take an alternative view of life.

Sometimes other birds move into a woodpecker space, such as great tits, blue tits, coal tits or redstarts. Starlings move in occasionally, although they can be less keen to wait their turn on the housing list and sometimes employ henchman-like tactics, ousting our poor hard-working woodpeckers before they're ready to move on, and squatting the place. To add insult to injury the starling has been known to use its fantastic mimicry skills to imitate the calls of woodpeckers.

CRUCIAL CREEPIES

Creep a step up or down the food chain – or better still, weave your way in and out of it in a carefree nature ramble – and the links that connect the oak's world start becoming more visible, sometimes literally visible to the eye, and sometimes coming to light in our understanding.

You start to become aware of the smaller things, arguably low, or at least lower down in the food chain, that hang out in oak trees. Bugs, creepy-crawlies, mini-beasts, call them what you will, they might, on first sight, seem lowly but they are highly important; not least as they are building blocks in an ecosystem. And what they – arguably – lack in size they make up for in numbers, with 1,178 invertebrates supported in one way or another by our oak trees.

Take the green oak tortrix moth (*Tortrix viridana*) for example, also known as the oak leaf roller. Its caterpillars are leaf eaters par excellence. They consume vast qualities of our oak's lovely, wavy leaves, earlobes and all, and can cause dramatic defoliation in the process. Add in other oak leaf munchers such as the winter moth (*Operophtera brumata*) and the mottled umber (*Erannis defoliaria*) and our oaks are sometimes stripped to the point where the tree is almost bare.

Thankfully for the oak, the caterpillars also have a copious capacity for being eaten themselves, as they are favourite foods of our feathery friends. Birds such as blue tits and our pied flycatchers will munch on caterpillars and, in feeding themselves and their families, perform a vital service for their home plant – the oak tree itself – by keeping the moth populations in check.

As well as this help from its feathery friends, the oak has other strategies to recover from caterpillar-induced leaf loss. As the spring progresses and melts into summer, oak leaves become increasingly tannic, which makes them bitter and harder to swallow for the tiny feasting beasties. And they have a little bit of magic up their branchy sleeves. A healthy oak can pop out a replacement set of leaves, sometimes called summer shoots but more poetically termed Lammas leaves, the latter coming from the old festival Lammas Day or Loaf Mass Day, which takes place on 1 August to celebrate the harvest of grains. This takes its toll of course. Experiments have shown that, while young oaks can recover from defoliation sometimes even more than once in a calendar year (for example, first by late frost and second by caterpillars), the effort required for the regrowing process reduces the ability of the tree to put on woody growth.[19] This makes caterpillar-chomping birds such as woodpeckers all the more important because, by keeping the moth population in check, they help support the oak's efforts to grow big and strong.

I can recommend taking a walk among some oak trees around Lammas Day to see if you can enjoy the sight of young, fresh leaves for yourself. It's a grand way of tuning in to nature and thinking a little about the old traditional Lammas festival too.

Each of the species that lives in, on or around the oak is a bit like a raindrop on a spider's web. Each offers a lens; its own unique view into oak life. We just need the light by which to view it. And, for me, each new bit of knowledge feels like a new ray of sunshine into this world.

PREDATORS AND PREY

Looping out into the wider web of life, the game of predator and prey intensifies; as the farmer and writer John Lewis-Stempel says, 'If our oak houses dedicated and exclusive species, it also harbours opportunists... A weasel can climb an oak, and into a green woodpecker's nest to eat the chicks', and, in a Winnie-the-Pooh-like twist, 'it may gorge so extensively that it is unable to exit until it has slept and thinned.'[20]

That woodpecker eggs – and those of other wild birds – are on the menu of other creatures is difficult to contemplate if, like me, you have fallen a little in love with woodpeckers. Perhaps even more difficult is acknowledging that our beloved woodpeckers themselves aren't always perfect angels. (In their defence, though, who is?) The great spots sometimes predate on their smaller, more vulnerable cousins, lesser spots, even going as far as to drill an extra hole in a tree to steal the babies or the eggs. This is all the more heartbreaking given the lesser spot's precarious hold on life: it is on the UK Birds of Conservation Concern Red List. The main causes of their decline are debated but the loss of mature woodland is likely to be a key factor.[21]

OPENING A DOORWAY TO NEW SOUNDS

The thing about nature is that you set off on one journey of discovery and find yourself enjoying multiple side trips along the way. That is part of the joy. My great spotted woodpecker opened up a whole birding world to me. It was a kind of gateway bird to a bigger birding habit.

I mentioned I got a bit of a lift, a birdy buzz, on hearing my first woodpecker. Truth be told, it was more than just a bit of a buzz, it was a buzz surprising in its extent. It was a lovely little life high, satisfying and exciting at the same time, like the moment when you land a ball neatly in the back of a football net or suddenly realise the answer to a difficult crossword clue or hit a difficult note just perfectly in a song. I have since heard other people describe similar feelings. In the podcast *Get Birding*, actor Samuel West describes his first time of being able to identify a bird by its sound, in this case by its song: 'I remember the first time I was walking with my girlfriend in, I think it was Epping Forest, and we heard a very, very high *tse tse trrrr trrrr*. I said to her, "I think that's a long-tailed tit", and I went round the tree and it was! And I felt like a superhero, I was so excited, I said to her, "I've got these bionic ears, I don't need to look, even", and it was a completely captivating feeling.'[22]

For me, identifying my first bird sounds opened a new door into life, particularly into the life of the oak and its wildlife. That first success set me on an extra, unplanned journey, a journey to learn bird sounds. Perhaps more importantly, it opened my ears. It got me tuning in to other sounds around me and made me realise how little attention I paid to my ears, and to listening in general: I was blissfully taking for granted this wonderful ability that most of us are lucky enough to possess. I felt glad to have rediscovered the power of this sense and grateful that I could use it more. Best of all, when it comes to buzzes, to little life highs, it is a free one, one anyone lucky enough to have reasonable hearing – and with apologies to those who haven't – can achieve in the right place at the right time. I appreciated this opportunity to make the most of my senses and to stretch my listening skills.

But it wasn't easy. I had tried to learn about bird sounds several times before and had kind of given up. I have come across many other people who have felt the same, even people who were dedicated listeners to BBC Radio 4's *Tweet of the Day*. It is easy to get lost in the beautiful aural blur of tweets and flutes and trills.

This all changed for me at the start of my oak adventure thanks to another guide, this one an old friend, Julian Branscombe, chief executive of the Isles of Scilly Wildlife Trust. Julian shared some great tips, but one particularly brilliant one proved to be the key to the doorway of nature's sounds for me.

The advice that Julian gave me that changed everything was to listen to one bird only, a bird like a robin, one that is common and that you can easily identify by sight. Don't try to listen to anything else. And really listen to your chosen bird. Keep at it.

I did as I was told, not with any great hope of success but more because I had agreed I would and I didn't want the guilt of letting Julian down when he had been generous with his time. Luckily I have access to robins in the garden I share with husband Toby, Pepe the pooch and the wild things, as well as on the paths around our village. I listened and listened and stood beneath one little fella as he sang his little red-breasted heart out in a tree that overhangs our garden. And it was all very pleasant.

And then it happened. One day I was standing in our kitchen making a cup of tea and I heard a creature singing so loudly and piercingly that the sound cut through walls and windows. My first thought was, 'Hang on, that's not a robin.' I shot out of the kitchen and there up in the tree was – behold, you beauty – a song thrush.

This little birdsong achievement gave me energy and enthusiasm. I was on my way to learning birdsong. I messaged Julian in my jubilance, and he gave me some song thrush listening tips. They often repeat their phrases like they are quite pleased with themselves, as they should be because they are excellent songsters.

The other great tip that Julian had shared at the outset was 'start now'. This was a particularly helpful tip because it was February and there are only a few choice birds singing away in February, so it is easier to get your ear in, like trying to make out the sounds in a trio of instruments as opposed to an orchestra. Robins sing all year round in the UK but a great many birds won't get going – or even get over here from their trips abroad – until the spring. And it is especially lovely to enjoy the birds when you can see them all the better for the unleafiness of the trees.

... Learning to Recognise ... Bird Songs: Top Tips

- Start in late winter if you can. It is a good time to use both eyes and ears as you stand a good chance of seeing birds while the branches are still bare. 'Flirty February' is perfect as our feathery friends tune up for a spot of bird-style seduction.
- Try Julian's top tip and focus on one type of bird to begin with, one that is often found near you – in my case that robin – and don't try to listen to anything else for a while. Really listen until you feel confident you have that one song nailed.
- Focus on a small group of bird species that you know are in the area. If you jot down a list you can tune in your ears with some online listening.
- Your repertoire might include blackbird, song thrush and mistle thrush because they all sing early in the year. They can be confusing because their songs have similarities but I came across a great tip on social media shared by @Lucy_Lapwing who credits Dominic Couzens for coming up with it.[23] He said imagine

the bird is a lecturer. A blackbird will sing something then pause (long enough to write it down). A song thrush will helpfully repeat phrases several times. A mistle thrush is a less helpful 'lecturer' – like a blackbird, but much faster, without the pauses.
- Build up your list slowly. Next perhaps add a wren, which has an incredibly loud song – it has been compared to a teeny machine gun, and then a dunnock, which sings from prominent posts – perhaps atop a hedge – with an energetic, fast, squeaky song. By mid March you'll hopefully be adding other birds to your repertoire, as by this time of year all of our resident bird species are in full song, whilst this is also the month you'll hear the first summer visitor, which is likely to be the chiffchaff and which – helpfully – sings out its own name.
- Explore some of the great resources that are out there: invest in a birdsong app or CD or listen to BBC Radio 4's *Tweet of the Day* or the website xeno-canto.org. *Xeno canto* means, charmingly, 'strange sound'; it is all about the noises birds make and is the basis for an online community where people share their recordings and their expertise: citizen science at its best.
- Expect all this tuning in to new sounds to make you feel tired. I realised I returned from walks with Pepe the pooch more tired than normal after I began to listen, really listen. I think this is a good sign, as it's – presumably – a result of exercising different faculties and using different parts of your brain.

· · · · · ·

As I found out when I first heard a woodpecker drumming, bird sounds aren't all about song, of course. They produce other nice noises to enjoy and which will tune your ears in to nature. You can sometimes tune in to foraging taps, and then there is of course that woodpecker drumming. A good old hammering can ward off pesky rivals and attract a mate.

In *Woodpeckers*, Gerard Gorman explains that the drumming of the great spotted woodpecker builds up from the early spring until males might tap away hundreds of times a day. Their 'explosive bursts' can result in 10–16 strikes in only a second and this can be repeated several times in a minute. The females join in, but to a lesser extent.[24] The more resonant the wood, the better – those great, old, dead, skyward-reaching wood branches in trees known as 'staghorn oaks' can be perfect. These are old oaks in which the tree has started to save resources by reducing its canopy (more on this later). I have heard that a prime drum in resonant wood means the sound can carry up to half a mile.

The great spot is the most hard-hitting of the UK woodpeckers. The little, rare, sparrow-sized lesser spotted woodpecker is more your machine-gun drummer to the great spot's alpha showman, taking its time with a longer, more even tempo, sometimes with a touch of a rattle caused by the lesser spot's choice of instrument, which might be a thin branch. The third and last (if you don't count the occasional visit of cousin wryneck) is the green woodpecker, our Laughing Betsey. This woodpecker also drums a little, but not so much, and in short, soft spurts so we are less likely to hear it; drumming seems to be more effort for our green friend. We are more likely to pick it out of the soundscape through its comedic qualities, the fantastic 'yaffle', that mocking laugh.

PEOPLE IN TREES

So there are birds, beetles and caterpillars living in and around oak trees, alongside all kinds of other beings. But what about human beings? Robin Hood and his merry band were said to seek cover in the great hollow belly of an oak tree. Could that be true? Could an oak really house a merry band? Can oaks home humans, as they do nuthatches and moths?

I revisited my journey 'past The Bee Tree, up to The Major Oak...', later on, this time looking through a different lens, that of humans hanging out in oaks, and that brought me to The Major Oak.

The Major Oak is celebrated with good reason. It is a great, majestic giant of a pedunculate oak standing proud in the heart of Sherwood Forest with a 10.66-metre (about 35-foot) circumference and a canopy stretching 28 metres across (about 90 feet).[25] Standing in front of this beast with its whopping great hollowed-out belly, I heard a man tell his children that this is where Robin Hood lived with his merry men. It is not an uncommon belief, and it is a great story. It is easy to imagine a group of fellas, clad in Lincoln green – the city of Lincoln is just down the road – and their famous tights, carousing in and around this great tree. All of them would be obligatorily merry on account of their historical branding and possibly due to a spot of mead: sweet, sweet, fermented honey from nearby bee hollows in another famous oak tree, The Bee Tree.

It is also easy to picture Maid Marian green-queening it among them in this spot, although, personally, I'm not entirely convinced she'd have been thrilled to share a tree hidey hole – wonderful as a prodigious oak hollow is – with a band of meady men. I can't help thinking that the celebrated merriness might have worn off when the mead ran out.

Anyway, to be honest, if all or any of this happened – and that's a big 'if', isn't it – it wasn't inside The Major Oak, not as it stands today anyway. It's not that The Major Oak hasn't been living here a long time – estimates of its age range from 800 years up to more than 1,000 years, so it could date back to the time of our Robin's enemy 'Bad' King John, who reigned around the year 1200. It's just that it would have been something different back in the day. In 1200 The Major Oak might have been about 200 years old: very healthy presumably, given the long life it was to lead, if a little youthful in oak years, so let's say solid but reasonably slender in oak terms. It could, possibly, have been part of an older tree, one whose parts had regenerated over the years, restarting life almost from scratch, a process that can make oaks immortal in a way (we'll look into the idea of tree immortality later). Or it might possibly have been somewhat younger: an acorn perhaps, or some earlier nascent thing, possibly a puff of pollen floating around on a breeze that could have wafted across the face of our Robin.

That is not to say that a medieval outlaw couldn't have lived in a similar tree back in the day. There would be worse places to hang out. And we know for certain that people have spent time in and around The Major Oak at various times. High-profile visitors include, for example, Emmeline Pankhurst around 1912 – we are told she was one of 30 suffragettes to climb inside the tree.[26] To digress slightly, a sessile oak planted in Glasgow's Kelvingrove Park by suffrage organisations on 20 April 1918 to commemorate the granting of votes to women was voted Scotland's Tree of the Year in 2015.[27] Many big, old oaks have served as venues for various events across the centuries. Whether they were the really old beasts with hollows the size of sheds in their bellies or whether they served as big leafy parasols, havens from the weather, oaks have been used as venues for weddings, for preaching, for makeshift parliaments, for fomenting rebellion, for playing in and just as somewhere notable to meet.[28]

I was once lucky enough to meet a tree with a claim to be the biggest sessile oak in Britain (albeit one that is now hollowed out to the point that it is split into several separate parts – we'll look into how that happened later in the book). It is the mighty Marton Oak in Cheshire.[29] Over time it has given service as a pigsty, bullpen and playhouse.

The aptly named Tea Party Oak in Ickworth Park in Suffolk was once the gathering place of local children, who would meet beneath its branches on holidays.[30]

And The Owen Glendower Oak – which should perhaps be called 'Owain Glyndŵr's Oak' as Archie Miles points out in *The British Oak* – is in Shropshire, perhaps surprising given that it takes its name from the last native Welsh person to hold the title Prince of Wales.[31] It had 'sufficient room for at least half a dozen to take a snug dinner', back in 1810, according to D. Parker in correspondence to *The Gentleman's Magazine*.[32]

The Parliament Oak in Sherwood, which is possibly older than The Major Oak – some estimates go as far back as 1,200 years – is said to have sheltered King John of England as he held parliament, back in 1212. This old beast is even more hollowed out than The Marton Oak – in fact, it is more hollow than trunk.

Other oaks helped nurture seeds of rebellion by virtue of the shelter they provided. In 1549 Robert Kett led a peasant army revolt against the enclosure of common lands in Norfolk. Several gatherings took place under oak trees, resulting in various trees being dubbed 'Kett's Oak'. One, also known as The Oak of Reformation, made for a leafy HQ on Mousehold Heath, where Kett's men set up camp. This tree was – reportedly – chopped down to make way for a car park in the 1960s.[33] I wonder what Joni Mitchell would have made of that. Today there are claims for other Kett's Oaks that are still living, including one near Hethersett.

The bigger and older the oak, the more likely it is to be a venue for meetings of people, whether inside its belly or beneath its branches. However, while fun, frivolity and the stirrings of rebellion took place in and under oak trees in times gone by, we know better than to do this now. In recent years we have learned that trees suffer from the impact of human feet (we'll revisit this issue later in the book), so our contact should be delicate and respectful, especially towards the most precious and ancient of them. This doesn't mean we can't have a tree party, though. We can still enjoy being out in woods and the open, but let's tread lightly around our trees and let's protect our most precious oaken treasures by keeping a careful distance.

THE MAGIC FOREST
On my next pandemic journey into Sherwood I retraced my steps 'go past The Bee Tree', etc. but I didn't get beyond The Bee Tree, not for a while, anyway, because this tree is worthy of a linger. It is a gnarly old beast, all burrs and burliness, named for housing wild honey bees in its cavity. In summer it would be worthy of an even longer linger because there is a helpful post nearby with a spyhole and an invitation to peer through it. The hole is trained on the bee hollow, so that when our busy friends are buzzing about you can see exactly where to look for them. It is one of a great parade of ancient oak characters in Sherwood, including Stumpy, Moaning Mabel, The Man-eating Caterpillar Tree[34] – which is said to swallow anyone who enters the great doorway of its cavernous hollow – and The Medusa Oak, whom we'll meet later in the book. We are so lucky in the UK to have

historic tree personalities dotted across our landscape. Those I've just named are only a handful of Sherwood's characters. If you are already chomping at the bit to meet some of these wonderful trees, I recommend the Ancient Tree Inventory website and the map in Archie Miles' book *The British Oak*, where you can read many more oak stories.[35]

As I was leaving Sherwood Forest that day, I stopped to watch some nuthatches that were hanging around the entrance to the forest like a gathering of goths. Nearby was a thrush looking all retro with its speckled chest; it reminded me of someone wearing an oatmealy knitted tank top. The place felt full of fun and surprise and character. A woman, walking alone, was coming towards me. I nodded in the way strangers greet each other, but she spoke to me as if we were old friends in the middle of a conversation. Shiny-eyed, she said excitedly, 'actually I have just heard a woodpecker', with no introduction or context. Neither was needed. I understood. We agreed it was magical and we walked on.

2

Wasps, Words and Oakspirations

GALL WASPS ARE THE
ROCK STARS OF THE GALL WORLD
JAMES HEAL

THE WASP FACTORY

There is a little brown wasp, so small you have probably never noticed it. It is nothing like your bold yellow and black wasp that hangs out in gardens in summer, hoping for a go at your ice-cream. It won't even think about stinging you – it isn't that kind of wasp. It looks modest, humble, insignificant, yet this tiny, oak-dwelling creature, only 1.5–2 millimetres long (about $^1/_{16}$ of an inch), in many ways changed our lives. It boosted our culture and had a wide-ranging impact on the rest of the world.

So who is this mysterious creature? It is the marble gall wasp. Its home – the gall it makes – was once used to make ink, and that is how the wasp affected us so deeply. Oak gall ink – or iron gall ink – was used to write many of the most important documents in western history, from medieval times right up until the twentieth century: the Domesday Book; the Magna Carta; Shakespeare's plays; Charles Darwin's letters; Isaac Newton's theories; the confession of Guy Fawkes and, on 4th July 1776, the declaration of America's independence from Britain...[1] The list goes on.

This ink is one of many oaky links with our written culture. It feels somehow fitting that oaks have an indelible connection with writing, because the tree and some of the species it houses have inspired great works of literature. The oak inhabits both poems and prose, and we will dip into some of these later in this chapter.

I first saw a marble gall during a walk in Somerset with a couple called James and Alka Hughes-Hallett. They have planted many trees – a mixture of species including oaks – to create new woodland, an important personal contribution to the future when you consider how woefully under-wooded the UK is. The Hughes-Halletts' young trees – only a handful of years old – stretched out and away from us into the distance like a battalion of leafy little climate soldiers. A chiffchaff was singing its see-saw song, apparently celebrating the end of a rain shower, when James stopped suddenly, leant down and picked something from the thin branch of a baby oak that would have been perhaps three or four years old.

He straightened it up and handed me a smooth, rounded thing. It looked like a woody, brown marble. 'I collect these and give them to people,' said James. I rolled it in my palm: a marble gall, former home of its marble gall wasp. It had a tiny, woodworm-like hole where the wasp had vacated its property. It was the best kind of present because it came with a story.

The marble gall wasp (*Andricus kollari*) was introduced from the Middle East in about the 1840s, so it isn't native to the UK but it is widespread now. It is an incredible creature that gets our oak trees to do its bidding and become a nursery for its babies. It is a 'gall causer': it lays an egg in an oak bud; it uses chemicals that compel the tree to produce a gall, this marble-like sphere. This provides a home for the little wasp larva.

More than a thousand years ago, someone discovered you could make ink out of these galls by mixing them with a bit of iron sulphate, water and gum arabic (which is made using certain kinds of sap). And then the rest is – pretty much literally – history; a wonderfully tangible kind of history. We have now hundreds of years' worth of iconic documents in museums and places like the National Archives and the British Library, all written in waspy gall ink.[2]

That's not to say that all the ink came from UK wasps. William Bryant Logan says the best galls for ink came from an oak that grows around Aleppo, in his book *Oak: The Frame of Civilization*. Wherever you sourced your ink, it held great power. As entomologist George McGavin says in BBC Four's *Oak Tree: Nature's Greatest Survivor*, 'Everyone from poets, musicians and mathematicians to fine artists used this ink to record their thoughts, feelings and ideas... unwittingly the oak tree has enabled us to record our past, to express our most profound ideas and to share our deepest emotions.'

To humankind, the marble gall was a world changer, but to the wasp's grub a gall is home, and a des res at that, coming fully furnished with everything the youngster needs, including dinner. The grub munches the flesh of the gall; cosy and yummy at once. It grows up and emerges as a wasp (around September or October in the south of the UK, and in the spring further north), leaving a little tunnel behind it – that hole we mentioned earlier. There is a video on YouTube showing a wasp emerge – excellent wasp watching.[3]

Now, here's a strange thing. The wasps of this generation that grew up on our native oaks are all female, every one of them. Another odd thing is that our gall girls go on to reproduce without having sex. They lay eggs in the buds of Turkey oaks, or *Quercus cerris*.[4] These oaks were introduced to the UK in the eighteenth century, and although non-native to the UK, they help complete the life cycle of our wasp, producing a different kind of gall. The eggs of this generation hatch into grubs both male and female. On growing up they get together to mate and then make our marble gall. And so there is this strange cycle of sexual and asexual generations.[5]

The weird world of the marble gall doesn't stop there. Delve further and you find out that there is a lot going on in this tiny universe. It opens up more new doors to nature than I ever imagined existed. To understand more I needed a new guide and this is how I came to meet James Heal. James is a 'galler', someone who goes out hunting for galls on plants and studies them. He modestly eschews the fancier title of 'cecidologist', despite his impressive knowledge. I first met James when he gave a webinar on galls for the London Natural History Society (LNHS). This was during a period of Covid lockdown, and the LNHS was one of the wonderful organisations keeping us entertained and helping us by sharing a window onto the world beyond our homes. Afterwards he kindly agreed to meet online to tell me more.

James told me that inside the marble gall, things are so pleasant, what with all that food and cosiness, that it can attract guests, some more sinister than others. Some of these visitors are called insect inquilines. They 'exploit the living space of another, but there's a community in the space', says James. So, basically, it's a kind of insect flatmate thing, albeit the kind who of flatmate raids your food stores, so is probably a bit annoying, but won't do any harm. These inquilines include other kinds of wasp, such as *Synergus gallaepomiformis* that, as far as I can make out, has no common name, which seems a bit of a shame. Perhaps we should take inspiration from Shakespeare, who liked to compose new names for things, and make one up – how about the roomie wasp?

Then there are the less welcome guests. In fact they are more like gatecrashers. A growing gall wasp can attract its own parasite insects, the parasitoids. If the gall wasps are pop stars, the parasitoids are your troublemakers, living off their hosts like uninvited thugs. There you are, having a bit of a gall party, all reasonably pleasant with the mild exception of guests who are being mildly annoying by scoffing too many of your nice nibbles, but then a bunch of real undesirables turn up in their bovver boots. They aren't here for pleasantries, they cause damage. One of these, *Torymus nitens*, is also a kind of wasp; a 'chalcid wasp' in this case.

I was beginning to dislike this little wasp because of its impact on my new favourite wasp, until I learned it is one of those species that are highly dependent on the oak tree for their very existence. At heart this troublemaker is quite fragile in its own way; it can't just turn to any old plant to make a home, so it deserves a little love too. So, anyway, back in the marble gall, things can get pretty full as newcomers crowd in. In all there might be 50 insects found in a single marble gall. 'If you cut open a marble gall,' says James, 'you would find multiple different tunnels and holes of different sizes. If one tunnel is a different size to another it will be a different species.'

Even fungi might join the party too; sometimes you can see dark brown growths on older galls, revealing the presence of the fungus *Phoma gallorum*. (More on fungi later.)

Some unfortunate creatures even get trapped in the gall, hemmed in by the crowd like nightclubbers who have entered an interesting spot but can't find the exit. James continues: 'Once when I cut open a gall this spider came out. It was occupying some of the space of the gall, possibly just called in for shelter when it was raining and might have grown too big.' By way of an aside, James tells me he was only able to identify this spider by examining its genitalia under a microscope. It turned out to be a *Clubiona* spider. I ask James if the spider was still alive when he released it. 'Yes, it crawled out into my office and disappeared; it is probably still there to this day.'

But I digress. (I'm finding digressions are an enjoyable, almost inherent part of learning about oak trees.) Back to the galls. It turns out there are thousands of different kinds of gall. And they come in all sorts of shapes. While our marble gall is a beautiful sphere, some look like purses, some resemble helmets and others are as spiky as a pincushion.

While galls can be found on many plants, the oak has a special place in gall world; 'Some of the most complex, fascinating, beautiful galls are found on the oak,' says James. 'The oak basically breaks every record in terms of biodiversity in the UK. It supports more gall causers than any other plant. It also hosts a larger proportion of some of the most complex galls of any plant host, some of which we don't even understand yet.' Cynipid wasps – our marble gall wasp is one of them – make up a large number of the gall causers. James points out we should be grateful to these wasps because they introduce a lot of people to cecidology, 'If you don't find these things beautiful there is something wrong with you. Gall wasps are the rock stars of the gall world.'

While wasps are the rock stars, other creatures also cause galls. 'There are psyllid bugs [also known as jumping plant lice] and gall midges among other causers,' says James. Galls caused by the midge *Macrodiplosis pustularis* are easy to find. Look at oak leaves when they are relatively developed and often you will find the edges are folded very neatly. They are flat, like they have been folded with a ruler. The leaf is bent over specifically to provide sustenance and shelter for the gall midge larvae inside. At this point James reminds me that, although it looks very much as if someone has folded the leaf by hand, the changes in the leaf were caused by chemistry. 'That is all part of being a gall – galls are brought into life by the chemicals emitted by the gall causer; there is nothing mechanical about making a gall. This is all part of their magic.'

Another gall that is fairly easy to find is that of *Trioza remota*, whose adult looks a bit like a fly to me, although it is actually one of the 'true bugs'. The Amateur Entomologists' Society explains, 'The Hemiptera are called "true" bugs because everyone – entomologists included – tend to call all insects "bugs". That is a loose term, whereas the true bugs are just those contained within the insect order Hemiptera.'[6] This large group of insects accounts for around 75,000 species worldwide – we should note that estimates can vary. They all have piercing mouthparts, with which they can suck the juices from plants or animals. *Trioza remota* galls are common and easily found, says James. 'On the upper side of the leaf are dimples. Underneath there is a depression and filling it is the nymph of the bug. It spends a significant proportion of its life in that period, using the leaf for sustenance and protection.'

It took me a while to get my own galler's eye in, and then – in a similar way to my birdsong breakthrough – I suddenly got it. It was a sunny May day and I was out with Pepe the pooch on a country path near our home called Hopyard Lane, which is lined with oaks of all ages. Their leaves were still spring-small, the trees themselves not yet too crowded. I was rifling through a low-hanging oak branch on a youngish tree of probably 200 years – The Tree House Oak, I call it – and I spotted a tiny sphere, not even a centimetre across. The little orb was pale with rosy pink patches and was cradled by a spring-bursting bud that looked like an upturned hand. This gall was an oak apple.

Oak apples aren't really apples at all, but they can do a good impression of a little, squishy one. They smell sweet, and people have been known to eat them, insects and all. They even once had their own special day: Oak Apple Day, on 29th May, which was related to the restoration of the monarchy and the oak's special role in that (more on that later).[7] The oak apple is created by the wasp *Biorhiza pallida*, which is another of James's rock stars. It is found on buds in early summer and is a great 'starter gall', says James, something to get your galler's eye in with and a good one to look for with children because it is easy to spot and identify.

Galling turns out to be a way of getting up close and personal with the oak and its world because it gets you among the leaves and the low-hanging branches. It is a bit like playing 'Where's Wally?', except this is more 'Where's Gally?' It is a fun thing to do with family or friends. Together you can explore many diminutive worlds and wonder at the beings they house.

Now I seem to see galls everywhere I go and it feels strange to me that I never noticed them in the past.

· · · Getting into Galling · · ·

I asked James for some tips to share with people who might like to have a go at galling. He said April to September is the main season for galls: 'If you are looking at oak leaves in summer you almost can't fail to find galls… similarly, if you are looking at the acorns later on.' I would add that it's good to find an oak or two with low-hanging branches so you can get up close and have a good look at the leaves and twigs.

IN SPRING, LOOK FOR THE FOLLOWING:

Oak apples: This is James's 'starter gall', the apple-like gall that can be found on oak twigs, caused by the oak apple wasp. You are less likely to see the other version, which is made on the oak's roots.

Spangle galls: These are lovely disc-shaped growths, some brightly coloured. The common spangle gall is caused by the spangle gall wasp, *Neuroterus quercusbaccarum*, and can be found on the undersides of oak leaves. A single leaf can host up to 100 galls, each containing a single larva. The larvae emerge as adults in the spring. The next generation develops in oak catkins or leaves, creating a different kind of gall, a round, version that dangles and looks like currants.

IN SUMMER, LOOK FOR:
Knopper galls: These make acorns look like they are wearing a knobbly helmet. The name is related to the German word *knoppe*, which, indeed, refers to a kind of helmet. They are sticky and red to start with, later becoming woody and brown. You find them on the acorns of the pedunculate oak. Their causer, another wasp, is called *Andricus quercuscalicis*. A second generation then develops in the catkins of the Turkey oak.

Oak marble gall: These are our hard, woody, marble-like balls that can be found on oak twigs, often in clusters, caused by the wasp featured above. They are often on young trees. They turn brown as they mature and emergence holes, from which the asexual adults have escaped, can be seen from autumn onwards.

Hedgehog galls: These are harder to find but full of character, a treasure if you are lucky enough to spy one. The sticky, spiky gall is caused, appropriately, by the hedgehog gall wasp (*Andricus lucidus*). Again this is one of two generations of galls.

IN LATE SUMMER, LOOK FOR:
Silk button galls: The silk button gall wasps that make these spanglers have a lovely scientific name, *Neuroterus numismalis*, in that *numismalis* means coinage and they are sprinkled across oak leaves like a collection of tiny gold coins. James says, 'The life cycle of this species – another wasp – involves making a more subtle gall; it is a similar shade of green to the leaf and makes a

slight blister. Adults emerge from blister galls and oviposit – lay eggs – in the leaf and silk buttons emerge from the leaves. They stay on the leaf or under the ground over winter.'

IDENTIFYING YOUR GALL

Once you have found your gall, here are the three key steps to help you identify it:

1. Identify the host plant – e.g. oak, but different plants are, of course, a source of galls too. This step is often forgotten – a classic beginner's error – but important in helping you narrow down the type of gall you've found.
2. Magnify, photograph and/or collect your gall. Many plant galls can be identified out in the field, and it can be helpful to have a device that will help you magnify your gall. James recommends a phone lens clip. Magnifiers on smartphones can also be useful – try the apps library or the accessibility settings.
3. Consult a guidebook. For beginners, James recommends Michael Chinery's book *Britain's Plant Galls: A Photographic Guide*. There is also *British Plant Galls*, by Margaret Redfern and Peter Shirley, or the website Plant Parasites of Europe: Leafminers, Galls and Fungi.[8] For a quick view of some of the most common galls, keep handy Redfern and Shirley's Field Studies Council fold-out chart *A Guide to Plant Galls in Britain*.

· · · · · ·

Nowadays we rarely use galls for ink, although in 2017 gall ink was used to write the new Charter for Trees, Woods and People – Tree Charter for short. This sits in Lincoln Castle alongside one of only two remaining copies of the 1217 Charter of the Forest, which was signed alongside the revised Magna Carta by King Henry III and which was, says the Institute of Chartered Foresters (ICF), 'perhaps the first attempt to set down the rights of citizens to the sustainable benefits of the landscape'. The new Tree Charter was crafted by more than 70 organisations working together. The ICF says, 'It exists to demonstrate the huge groundswell of love for trees that exists under the surface of UK society. It also serves as a reminder to those who have become "tree blind" that there are many reasons to fall in love with trees.'[9]

MEANWHILE GALLS continue to play their role in nature. Zoom back out into the wider world and you find that galls help link up the web of life by providing food for other species. Mature galls are broken open by creatures who are after the grub – you can sometimes see the leftovers of an oak apple meal after robins and blue tits have torn it to bits to winkle out the goodies from the inside. Woodpeckers and other creatures such as voles and mice also sometimes tuck in for lunch.

We humans have found other uses for galls too, especially the oak marble gall. I understand you can make a decent deodorant from them due to their antibacterial qualities. They are also said to be an excellent cure for piles if mixed with hog's lard and applied to the posterior. Let's face it, sometimes in life you need a book, and other times, relief from an uncomfortable complaint.

BRANCHING OUT INTO LITERATURE
Many writers have expressed their thoughts using gall ink. And it is fitting that the oak's literary roots spread far and wide. The oak, it seems, is almost intrinsically inspirational. If you are a Roman poet and you want to express something kingly, what do you do? You write about the oak. If you're

a Victorian diarist and you want to convey majesty; or an Edwardian novelist and you are after a sense of safety and solidity, what do you do? Invoke the oak.

The Bowthorpe Oak
I once heard the author Jeanette Winterson, at the Hay Festival, point out that Shakespeare lived only half the life of an oak ago. Some of the individual trees that inspired people hundreds of years ago are still living. I find it mind-popping that the tree that moved John Clare to write his poem 'Burthorp Oak' well over a century ago is alive and well and living on a Lincolnshire farm.

The Burthorp Oak, now better known as The Bowthorpe Oak, can make a strong claim to be the English oak with the biggest girth alive today. What is sure is that it is an absolute whopper, measuring in at an immense 13 metres (about 42 feet). You can visit it today for a small fee to the farmer on whose land it sits.[10] If you go, don't forget to take along a copy of Clare's poem.

> *Burthorp Oak*
> Old noted oak! I saw thee in a mood
> Of vague indifference; and yet with me
> Thy memory, like thy fate, hath lingering stood
> For years, thou hermit, in the lonely sea
> Of grass that waves around thee! Solitude
> Paints not a lonelier picture to the view,
> Burthorp! than thy one melancholy tree
> Age-rent, and shattered to a stump. Yet new
> Leaves come upon each rift and broken limb
> With every spring; and Poesy's visions swim
> Around it, of old days and chivalry;

And desolate fancies bid the eyes grow dim
With feelings, that Earth's grandeur should decay,
And all its olden memories pass away.
JOHN CLARE

The other amazing thing about The Bowthorpe Oak is that it appears to be in better shape than it was at the time Clare wrote his poem. When Clare writes of an 'age-rent' tree 'shattered to a stump', he seems to describe a plant that is on its last legs, or – as we should probably say – its last roots. Well, it is certainly ancient. Some estimates of its age surpass a thousand years. Its trunk is stump-shaped and hollow, but when I visited it in mid May in 2021 it seemed to be thriving and offering a lesson in growing old gloriously. John Clare mentions that leaves come 'with every spring' and, even two centuries on, he's not wrong. The tree I saw was covered in a rich, lush spring coat of bright green foliage.

The loneliness in Clare's 'lonely sea of grass' might well have contributed to The Bowthorpe Oak's longevity, as we now know that oak trees need space and light to achieve their full potential (we'll return to this later). It has also probably benefitted from being in the care of one family for six generations. Richard Blanchard is the current custodian. With his sons he runs the farm on which The Bowthorpe Oak sits, and he tells me about his tree, speaking with enormous affection. He explains how it was sitting in a kind of parkland or wood pasture until his family started farming the land around it: 'When it came to us it was a park, the area was ripped up because of the war, but where the oak was placed it hadn't been subjected to any big machinery. It's got to have been in the right place to survive as long as it has. We try to keep it as natural as we can.'

John Clare was not the only person to write about this stunning tree. In *The British Oak* Archie Miles quotes a diary entry from 1763 that seems to be the earliest mention of The Bowthorpe Oak: 'Visited Bourne and stopped at Bowthorpe Park to see the celebrated oak-tree there. The lower part was used as a feeding-place for calves, the upper as a pigeon-house.' A later entry tells how the oak was 'neatly fitted up inside with tables and seats, and could contain a "tea-drinking" party of sixteen persons'. The hollowed-out interior bears witness to this. Parts of the fittings are still there: a nail where a table was once attached to the inside of the tree, the remains of a flagstone floor and some graffiti recording dates as early as 1866. Given the size of the cavity today, the number of guests at the tea party appears to have been a bit of an exaggeration, either that or they were 16 very small people – very small and probably very squashed people. Having said that, Richard points out how much the tree has altered, even in recent years: 'It amazes me how much it has changed in my lifetime. It reminds me of me. I was a lot slimmer before.' Weren't we all, Richard?

Richard tells us a story from his childhood. He points out a low gap in the trunk where a little door used to sit in the side of the oak tree. The gap is now about four feet high and a few inches wide. 'When I was young, we had a pony that got trapped in that hole; it was a small pony, but still a pony.' These days you'd struggle to get a goat through the same fissure, so it goes to show there is an awful lot of change over recent decades in the shape of this old, old tree. Thankfully the pony story ended well: the animal was released from the bole hole (the bole being the tree trunk) with a bit of help from some human friends.

The Bowthorpe Oak continues to inspire people like artist Mark Frith, who featured it in his A Legacy of Ancient Oaks exhibition. Of oak trees, he writes: 'It is humbling to spend long hours contemplating these magnificent veterans, contemplating their longevity, their individual characters, their strength and symbolism... At the same time I grieve for the uncertain future of these ancient trees, given the all too real and imminent threat posed to them by climate change... of which [humans] are an inextricable part.'[11]

Herne's Oak
The many writers who have paid homage to the oak include Shakespeare himself, he who lived only half the life of an oak ago. In *As You Like It* (Act IV, Scene 3), he captures the essence of an old, old tree; the type of oak whose topmost branches have given up bearing leaves and are now bare:

> Under an oak, whose boughs were moss'd with age,
> And high top bald with dry antiquity,
> A wretched ragged man, o'ergrown with hair,
> Lay sleeping on his back...

In *The Merry Wives of Windsor* (Act IV, Scene 4), Shakespeare features a ghostly, old oaky tale:

> There is an old tale goes that Herne the hunter,
> Sometime a keeper here in Windsor forest,
> Doth all the winter-time, at still midnight,
> Walk round about an oak, with great ragg'd horns;
> And there he blasts the tree and takes the cattle
> And makes milch-kine yield blood and shakes a chain
> In a most hideous and dreadful manner

The spooky tale turns to merriment – of sorts – when the wives set up Falstaff into dressing as Herne the Hunter to meet them beneath an oak tree in the forest. Poor Falstaff turns up complete with stag's horns strapped to his head. He is subjected to humiliation while children dressed up as fairies dance around the oak, pinching him as they go.

The tale of Herne the Hunter was said to be based on the life of a real man called Richard Horne – or Herne – and his tree – The Herne's Oak – was said to be a real tree.[12]

Herne, the tale goes, was one of the keepers of the ancient Windsor Forest during the reign of King Richard II back in the fourteenth century. I have read many versions of his tale. My favourite has Herne and the king as hunting companions. One day, our hero, Herne, rescued the king from an attack by a stag. Herne was injured in the process, but he was saved by a mysterious stranger, who appeared out of the blue and treated him with an interesting medical procedure. This involved taking the antlers from the dead stag and attaching them to Herne's head, where they somehow fused on, like some kind of magical antler graft. Herne recovered – I know, let's roll with it – and was rewarded by the king for his courage. All would have ended happily, had his fellow hunters not become jealous. They accused our stag-headed friend Mr Herne of crimes including poaching deer. The king, having initially stood by his friend, started to believe the accusations. It was all too much for our hero, Herne, who went and hanged himself from a large oak: our tree – we got there in the end.

But there is more. You would have thought the sorry story might have ended there. But events got odder still when the scheming huntsmen did a deal with the ghost of Herne – yes, he was a ghost now – and were compelled to ride alongside him forever in the forest in an unfruitful hunt as punishment for their lies. (It might be best to view this through medieval eyes, when an eternal hunt without a catch was a kind of misery for a high-society person when the poshest people considered venison 'the noble meat', an important part of life.) Everything came to a head when, during a walk in the park, the king – presumably pondering on the shortage of venison feasts – saw a bolt of lightning hit an enormous oak – our tree again – the one on which Herne had died. Through the smoke, the king laid eyes on a horned figure: Herne. Herne the ghost told the king that if he punished the traitorous huntsmen, the deer would return. The baddies were hanged on the same oak and the deer duly came back. Revenge was kind of served – in its cold old way – and Herne stayed on to haunt the forest.

Whatever your version of the tale, there was once an old tree called Herne's Oak. It was one of the many ancient oak treasures of Windsor Great Park. Of course we can't really tell if Herne's story has some truth in it, or if the tree named after him was the right one. Either way, while that particular tree appears to be long gone (I have heard of a newer replacement), we can still see its own ghost preserved in artworks, including an etching called *Herne's Oak, Windsor Park*, attributed to Samuel Ireland around 1799 and featuring Windsor Castle in the background. Even better, we can get a sense of the scenes Shakespeare describes by taking a visit to Windsor Great Park today, with its collection of ancient oaks. Go late in the day as the light is fading, and you will find some of them are suddenly gloriously spooky.

The Birnam Oak

Much further north, in Perthshire, are a couple of elderly tree neighbours with a different Shakespearean connection. The Birnam Oak, a sessile oak, lives alongside another venerable tree, The Birnam Sycamore.[13] These trees are said to be the last survivors of the forest that once straddled the hills along the River Tay, and which Shakespeare featured in *Macbeth* as Birnam Wood.

When Macbeth visits the famous witches, he is hoping for reassurance about the security of his position as king of Scotland. Instead, they make a prophecy that he will only be safe until Birnam Wood marches into battle against him (*Macbeth*, Act IV, Scene I):

> Macbeth shall never vanquished be until
> Great Birnam Wood to high Dunsinane hill
> Shall come against him.

According to Shakespeare, Macbeth listens to this prophecy and dismisses it. After all, 'Who can command the forest and make the trees pull their roots out of the earth?'

Unfortunately for Macbeth, he didn't see a trick coming (despite the fact that, at the time, he was in conversation with three witches and a set of ghostly apparitions). He was, of course, in for a whole lot of hubble, bubble, toil and trouble. It turned out that, while the whole forest couldn't get up and walk all by itself, it had another way of getting to Dunsinane Hill. Macbeth's enemies carried branches from Birnam Wood, which hid their movement from his soldiers. The wood moved after all, in a fashion. And Macbeth met his downfall.

In a new twist on the story, quotes about the Birnam trees appeared on social media accompanying footage of Ukrainian soldiers in wartime action. The soldiers had disguised themselves with branches, and looked like walking trees as they moved through the woods. It makes you wonder if Shakespeare's inspiration was based at least partly on the realities of soldiering. Some things don't seem to change so much.

And while I'm on the topic of life imitating art, the connections in the web of life surrounding the oak are sometimes mirrored in our literary heritage, with its inspirations and links and twists and turns. Just as Shakespeare was inspired by old stories, his work became a springboard for others. The Birnam Wood plot twist in *Macbeth*, for example, influenced the creation of a set of twentieth century literary icons: the glorious Ents in J. R. R. Tolkien's *The Lord of the Rings*. However, Tolkien's creations seem to have arisen more from irritation than inspiration, as Tolkien explains in a letter to the poet W. H. Auden: 'Their [the Ents'] part in the story is due, I think, to my bitter disappointment and disgust from schooldays with the shabby use made in Shakespeare of the coming of "Great Birnam wood to high Dunsinane hill": I longed to devise a setting in which the trees might really march to war.'[14]

In another twist, Shakespeare could conceivably have witnessed The Birnam Oak (sometimes also called Macbeth's Oak) himself. He is said to have visited the area at the invitation of King James VI of Scotland, along with a troupe of actors.[15] The Birnam Oak's age has been estimated at 600 years and if so it would have been a mature tree in the Bard's time. Perhaps he even touched its bark, and thought of words and worlds and imagined trees and tricks. I'm guessing he probably never imagined that one day another great writer would accuse him of 'shabby use' of walking trees and from that go on to stir the imaginations of generations with his amazing Ents.

TRACING OUR ROOTS

In the seventeenth century, the poet John Dryden translated the following, by the Roman poet Virgil in the *Georgics*.[16]

> ... Jove's own tree,
> That holds the woods in awful sovereignty,
> Requires a depth of lodging in the ground,
> And, next the lower skies, a bed profound.
> High as his topmost boughs to heaven ascend,
> So low his roots to hell's dominion tend;
> Therefore winds, nor winter's rage, o'erthrows
> His bulky body, but unmoved he grows.

While the oak's literary roots go back a long way, Virgil wasn't entirely right about trees' actual roots reaching down into hell, unless hell is only a metre or two from the surface because, as with other trees, an oak's roots are surprisingly shallow. Virgil can be forgiven though, because this wasn't understood until much later. The great storm of 1987 brought this home when it uprooted 15 million trees in the UK. Arboriculturist Tony Kirkham, formerly head of the arboretum at Kew Gardens, told me, 'What we realised from the storm was how shallow roots actually are, because, you know, everyone would draw a tree and if it was a cartoon tree the roots would replicate the top and we were looking at all these root plates and we were thinking, hold on, where are all the roots and we realised that the average depth of a tree root was about 800 millimetres and a meter [about three feet] is deep.'

What a tree's roots lack in depth they more than make up for in reach. They spread out way beyond the tree's canopy. This structure is an effective strategy for outwitting winds and 'winter's rage' as a forester once said to me. Imagine trying to break a tree's root by snapping it in half. Now imagine pulling it like a cracker – it is much harder to break that way.

Virgil is perhaps on safer ground on the subject of tree-time, going on to say, 'For length of ages lasts his happy reign, And lives of mortal men contend in vain.' This is a theme that comes up again and again throughout history. What is our threescore-years-and-ten (if we're lucky enough to live a long life) compared to the oak's renowned ability to grow for 300 years, live for another 300 years and then spend another 300 years dying? It is all relative, isn't it, and even a 900-year-old oak can look like a spring chicken compared to a 3,000-year-old yew tree.[17]

William Cowper addresses the time theme too in his great but unfinished poem, 'Yardley Oak':

> By thee I might correct, erroneous oft,
> The clock of history, facts and events
> Timing more punctual, unrecorded facts
> Recovering, and mis-stated setting right.
> Desperate attempt, till trees shall speak again!
>
> Time made thee what thou wast, king of the woods.
> And Time hath made thee what thou art, a cave
> For owls to roost in…[18]

Yardley Chase in Northamptonshire and Buckinghamshire was home to the oak tree that provided the inspiration for this poem. Once a hunting area – reflected in the name 'chase' – it remains an important woodland site for nature, providing a home for, among other things, 30 breeding butterfly species.[19]

AS THE OAK is enmeshed in our cultural life, it has also threaded its way in and out of our language. Oaky names are dotted through our language like trees around our landscapes. These range from the obviously oaky names like Sevenoaks and Oakthorpe, to the less evident. The Anglo-Saxon word *ac* relates to acorns and oaks. It wasn't until I was an adult that I learned my own family name, Acton, means oak farm or village. And it was only on reading Archie Miles' book *The British Oak* that I discovered that the 'eik' sound in the name of a nearby village called Eakring appears to mean a ring of oaks.

BRAND OAK
In *Far from the Madding Crowd*, Thomas Hardy calls one of his characters Gabriel Oak, and with that surname signals that his hero is solid, constant and reassuring. James Canton picks up this theme in his book *The Oak Papers*: 'We, as individual human beings, like Gabriel Oak, can seek to strengthen our own sense of worth, I think. We can be stronger, fairer on this earth, and help to home and house and care for the creatures of this world that live cheek by jowl beside us.'[20]

Just the one word 'oak' can summon up a multitude of impressions. What with all this beauty, strength and steadfastness associated with the oak tree, as well as the associated art and poetry, it is no surprise that marketing experts have got in on the oak act.

The British Isles are awash with oak symbols: brand experts bring on board oaky logos and names in the hope of harnessing a bit of this oaky value for their business, their political party, their charity.

My two personal favourites, organisations with oak leaf logos that I think live up to these values, are the Woodland

Trust and the National Trust. Full disclosure – I work for the Woodland Trust. I can also say, as an insider, that I am deeply proud of the organisation, its values and its cause. I spoke to Dr Darren Moorcroft, CEO of the Woodland Trust, about the organisation's logo. He told me, 'It is iconic as a native species of the UK because of what it delivers. The leaves themselves are universally recognisable. Also for the logo, by placing the two oak leaves together you get the pathway through the middle. That is really powerful symbolism, which sometimes gets lost, in terms of how we, from an organisational point of view, are opening up woods and trees to people. It is also a pathway into woods which are free to access. What the logo does in both its structure and its symbolism, there is a power to it that I think is a really strong one for our cause.'

LIFE LESSONS FROM THE OAK AND A TEENY, TINY WASP
Arcing back to the words of George McGavin and the marble gall wasp, that the oak has helped us 'record our past' and 'express our most profound ideas', it all goes to show how the wonderful web of the oak's world weaves into our own, and how a teeny, tiny creature has touched our history, our souls and our lives.

There are many small, modest creatures that most people have never seen, let alone heard of or recognised. We know that at least 1,178 invertebrates are known to be associated with the oak tree. Maybe one day one of them will change our world the way the marble gall wasp did. Maybe they hold the key to a new medicine or a better way of living. Maybe some of them will die out before they have the chance to change our world radically. Imagine if that had happened to the marble gall wasp. Maybe they'll have little impact on human lives and culture. All I'd say to this is that if there is a small thing bothering you with its buzz or its flutter one day, if your first instinct is to squash it, please think twice. It is so easy to slap at a creature that has found its way onto our arm or our table just because it is causing a mild irritation. Big, strong giants that we are in comparison, we can so easily react and thoughtlessly take a life, and what an incredible life it might be.

3
Crowning Glories

SING FOR THE OAK-TREE,
THE MONARCH OF THE WOOD:
SING FOR THE OAK-TREE,
THAT GROWETH GREEN AND GOOD ...
MARY HOWITT, FROM 'THE OAK-TREE'

ROYAL RELATIONS AND THE HUNT FOR THE ELUSIVE PURPLE HAIRSTREAK

If the oak is king of the forest then the jewel in its crown is the purple hairstreak butterfly (*Favonius quercus*). This is the elusive, twinkling butterfly that hangs out high, high, high up in the leaves of the oak tree and flashes like a gemstone on a sunny day. What makes this little stunner all the more precious is that it is dependent on our oak trees for its very existence.

It is also incredibly hard to spot; one of those creatures that makes you hunt for a glimpse of its beauty. Today I have been offered the possibility of a peep at short notice, so I have dropped all other plans. Lockdown is blessedly over for now and I have driven as fast as we are allowed across the whole of Nottinghamshire, then Derbyshire, in the blasting heat of a gloriously, surprisingly hot, British, July day. I am going to meet Ken Orpe. Ken is Derbyshire's official Butterfly Recorder, which, I believe, makes him the official butterfly king of the county. Ken is going to try tracking down a purple hairstreak with me.

It's not like I haven't tried to find a purple hairstreak before. My husband Toby, Pepe the pooch and I have spent several pleasant evenings this summer hanging out with oak trees, squinting up into their canopies through binoculars in the hope of spotting one. Unfortunately, even with binoculars, we're too far away to have any real chance of recognising such a small creature.

Other butterflies were more accommodating, coming down to our level so we could sneak up with our Field Studies Council (FSC) fold-out field guide at the ready to identify them on the spot. Not purple hairstreaks though; these are the Greta Garbo of butterflies, generally living high up in their distant, oaky haven.

They are, says Ken, 'top of the heap' when it comes to butterfly spotting.

Ken and his wife, Pat, are waiting for me at Kedleston Park, near Derby, under the deep shade of a group of old oaks, which are in the full flush of their dark green summer leaves. This is a blessedly cool spot after the long, hot journey, and Ken and Pat, like many people who love nature, exude serenity. Unfortunately I do not share this quality, not today anyway, which is 'one of those days'. I arrive hot and flustered and, on pulling up, somehow manage to lock myself in my own car and set off the car alarm. It isn't a great start to a nature ramble, a general principle being to try to avoid scaring the wild things, but Ken and Pat greet me warmly, kindly pretending not to notice my clumsiness while I try not to let on that I have also just caught my thumb in the car door.

We set off on a wooded walk. There are oaks – ancient, mature and young – all around, and sunny glades are dotted among them. The Kedleston estate is a wildlife haven and also a piece of heaven, a parkland created for the aristocratic Curzon family with the help of renowned architect Robert Adam and now cared for by the National Trust.[1]

Ken slows in front of a tree that is small in oak terms, perhaps 6 metres (20 feet) tall; about the height of our little cottage. It tends to be easier to see purple hairstreaks on this size of tree, explains Ken, a simple case of the tree being shortish and so the crown – which our purple beauties so love – is lower, nearer to our eyes.

I wonder, could this be the moment I finally get to see one of the UK's most mysterious butterflies? While we are waiting, hoping, Ken tells me about the purple hairstreaks' strange lifestyle.

While these butterflies are known for hanging out in the oak's crown, they aren't complete strangers to the base of the tree because they pupate (transform from caterpillars to butterflies) down in the soil below. There they hang out with ants. Butterfly Conservation says, 'They produce substances that are attractive to ants, and can make noises like ants and have been found inside ant nests, so they seem to have a relationship with ants.'[2] The ants seem to be important allies. Could they be little butterfly nursemaids? I haven't yet found the full answer to this underground mystery but I haven't given up investigating it either – please let me know if you have any insights. On hatching, the young butterflies rise up high into the tree for a bit of midsummer fun and feasting on honeydew and butterfly-style flirting. The resulting eggs look – gorgeously – like tiny sea urchins and are laid on plump oak buds. In spring they hatch into fantastic, hairy, ridged little creatures. Nourished by bursting oak buds, the larvae eat and moult three times, each time transforming into a new oddity, at one point looking a bit like a skinny woodlouse; becoming a butterfly can be a complex thing. When it is time for the caterpillars to pupate they descend back into the soil or moss, back down in ant world.

'Look!' Ken says, suddenly, his arm shooting upwards, 'I can see one! Look – that branch sticking out towards us!' I follow the direction of his pointing finger, my eyes flitting about, desperate not to miss the star of the show. I peer and stare and squint and think I have missed it and then, suddenly, I see it too, silver, flashing. 'It's like someone is dropping silver coins. It has probably been disturbed by a bee that has dropped on it,' says Ken. 'See its flight, it is jazzy, jerky, like a dance, not one of those flippy floppy flights.' He is spot on and I am in awe. We are looking at insect royalty. As it disappears from view, we slowly exhale. I beam at Ken and he grins back, saying, 'If you had 600 people coming down this path I'd be surprised if one of them spotted a purple hairstreak.'

We walk on, discussing ways to share this joy at seeing wild things. Ken tells me he has been working at the Young People's Forest, in Derbyshire.[3] This is music to my ears. It is a project I have been involved with, and the story behind it is a great one. The Woodland Trust acquired the site – a former open-cast coal mine – in 2019, supported by partners including #iWill, a movement that helps young people get involved in social action.[4] The partners invited teenagers to participate and – when they were ready – to take up leadership roles in the project. The young people, along with their local community, went on to create a brand new forest, planting 250,000 trees across 160 hectares (400 acres). Within it they have worked on creating great habitats for butterflies, by planting native oaks and by creating open, sunny spaces with nectar-rich plants for the wider butterfly population. That is where Ken came in, supporting the work with butterflies. It is a project full of hope for the future, for both the future of wildlife and the future of young people.

LOOKING INTO THE CROWN OF THE OAK TREE

Are purple hairstreaks rare – as they are often billed – or are they just not that interested in meeting us? I asked ecologist Dr Nick Littlewood, of Scotland's Rural College, that very question. 'I suspect purple hairstreaks are one of the most overlooked of the butterflies because they are the hardest to see,' he told me.

So it is possible that there are more purple hairstreaks than we have imagined? We know they are being recorded in new areas, including quite a few new sites in Scotland, which might indicate that they are extending their range in places. It is also possible that the increase in sightings is because more people might be out looking for them. Meanwhile there is evidence of declines in their populations, so we need to do everything we can to look after their homes on and around oak trees.[5]

To see a purple hairstreak, it helps to get someone to point you towards the right trees, those where purple hairstreak colonies have already been spotted. 'They are very active,' said Nick. 'Sunny evenings are a good time when they race around in courtship, then the incredible thing is they very often come back to rest on the same leaf each time.'

Our purple beauties do highlight a wider challenge: the harder a species is to see, the less likely it is to be reported. It might be because they are too high in the tree for most of us to see; it might be that they are buried beneath the soil; it might be that they are just generally elusive. They are a reminder that we shouldn't always take nature at face value. I wonder how many things we miss.

Talking of elusive, did you know that many of our moths love nothing better than an oak tree?

MOTHING TO SEE HERE

Why don't we love moths as much as we love butterflies? They are amazing, and some of their names are pure poetry. OK, pretty much every butterfly looks like a little flying work of art – what's not to adore there – but many moths are just as glamorous. What is it about moths that turns us off?

Well, for one thing, there is the image issue, that certain reputation for making holes in our favourite woollens. In fact this turns out to be largely unfair. A mere two common types of moth are reasonably likely to have a nibble at your jumper. Only two. That's not a lot of moths, not when you consider there are 2,500 types in the UK alone.[6] Organisations like Butterfly Conservation point this fact out again and again until they are as blue in the face as a... let's go with an Adonis blue butterfly. It is about time the world took this on board. Come on, people, it is time to ditch the moth cloth wrath.

Another challenge is that meeting some of our moth mates can be a bit tricky, partly because lots of them come out at night. And then there are so very many kinds of moth that it can feel a little overwhelming for the nature lover.

The oak houses, feeds and shelters some corkers – some amazing moth mates to help get us off the ground. Oak-lovers include the green oak tortrix, the winter moth and the mottled umber. They produce an absolute army of oak-creeping caterpillars in the spring. I have heard that at times they are present in great numbers, and that their droppings falling to the ground can sound like rain dripping through the trees. They munch on tender, young oak leaves, often leaving them looking holey, tattered and – well – moth-eaten, by the summer.

These very hungry caterpillars are very important though. They are vital food for baby birds and they appear on the oak scene just when they are most needed to feed hungry beaks. Having said that, it would be unfair to see them only from the perspective of our ravenous bird babies. They are great characters in their own right. The green oak tortrix, for example, is, as its name suggests, a very green thing, I mean really gorgeously green, just the right kind of green to hide on a fresh, young, spring oak leaf. It sometimes goes by the name the oak leaf roller: the caterpillars curl themselves up in oak leaves just as you would do if you wanted to hide inside a carpet. Safely hidden from foragers, they can munch away at their oak dinner in peace. If something should happen upon them, a pesky wood ant, for example, the caterpillars have a little trick up their leaf: they drop down from the tree like a tiny insect bungee jumper and then dangle from a silky thread until the danger is passed.

In case you were worried about our poor moth-eaten oak, if it is healthy it should produce a new flush of leaves, as we touched on earlier in the book. Lammas growth is so called because it happens around about the time of Lammas, the Celtic harvest-time festival, on 1 August. Meanwhile our hungry bird friends also do their bit for the oak, by keeping the caterpillars in check and keeping the whole oak ecosystem in balance.

Moth Night

It is July 2021, it is late, dark, and I'm off to meet some strangers in a wood in Derbyshire. This may not sound like a sensible plan for a Friday night out, but these strangers are bona fide members of the Long Eaton Natural History Society and tonight is their Moth Night, which is part of a wider national initiative led by organisations including Butterfly Conservation.[7]

They don't stay strangers for long. Marion, Derek, Margaret, Nigel and Patricia greet me at the Forbes Hole Nature Reserve, all ready to share their love of moths with a newcomer. They are carrying chairs, flasks and nibbles. Marion is making strange buzzing and clicking noises. It turns out she is holding a bat detector, the reason being that once you have set up your moth traps, you need to wait for a while for the moths to arrive; taking a bat walk around a moonlit nature reserve is an excellent way to fill this time. Marion leads our band on a winding route through shadowy trees and night-blacked ponds, occasionally calling out 'Common pipistrelle!' or 'Noctule!' in response to a whirr emitted from her gadget. She is accompanied by occasional calls from birds – the kissy calls of baby tawny owls and the cheeps of young swans. A train creaks by, a reminder that this is an urban nature reserve, a wild idyll amid houses and offices, shops and roads.

Eventually we wend our way back to the moth trap, which is sitting beneath a mature oak tree in full summer leaf. The trap has a blindingly bright bulb, which uplights the oak's leaves, painting them an emerald green against the night sky.

By the way, the word 'trap' might sound a bit nasty but this is a humane affair. The Skinner Trap is a wooden box with a central crossbar supporting the bulb holder and a rain guard, which protects the 125-watt mercury vapour lamp. Two large, angled pieces of Perspex deflect moths downwards into the cardboard egg-box packing. This is, basically, moth furniture – comfy places our insect friends like to settle for a while.

As we approach, I realise our box is full of fluttering beings. I turn to Marion in excitement. 'Caddisflies,' says Marion. 'Oh!' I say. Don't get me wrong, caddisflies have their merits, not least their elegant, long antennae, but moths they are not. So we sit around under the oak, chatting and hoping the evening will bring forth its creatures. 'It isn't the best night for moths,' says Marion after a while, a little dejected. The weather hasn't been great this year – we had heaps of rain just as they wanted to hatch into adults and feed on the nectar of flowers. It doesn't seem to have put off the mosquitoes, though, who are enthusiastically dining out on my neck.

We wait and talk and wait and talk. Not much happens on the moth front. And then, towards midnight, they arrive. En masse. It is worth the wait, moths of all shapes and sizes making for a glorious coming. There are big ones and little ones and really, really tiny ones. There are dull moths, fancy moths and peppery-looking moths. There are big, clumsy moths that bump into us, and moths draped in white, furry cloak-like outfits that twirl in the air like tiny fairies dancing in the night.

Some of them enter the light box and settle on their egg-box armchairs. We humans, guidebooks in hand, excitedly work out their identities. They are beautiful plume, peach blossom, scarce footman and the spectacle. Derek tells us that back in the day, a hundred or two hundred years ago, many moths were named by vicars who had a bit of time on their hands. Poetic, moth-loving vicars.

Marion tells us we're not allowed home until we see a hawk-moth. Derek says sometimes the team are there until the morning. I silently wonder if I'll last the course.

Eventually the big guys appear: the hawk-moths. If gall wasps are the rock stars of the cecidology world, the glam hawk-moths must be the Lady Gagas and the Jimi Hendrixes of lepidoptery life. An elephant hawk-moth arrives, resplendent in bright pink and orange. It has a trunk-like snout. Then comes a privet hawk-moth, striped pink and black, like a dandified army officer at a ball. And the poplar hawk-moth, which looks like a set of flying leaves. We are jubilant – each and every one of us – and now we are allowed to go home, although I don't want to any more. This is compelling. And it is a new take on the all-nighter.

In the end we count 162 moths and 57 different species. I say 'we' – it is Marion who logs the details and shares them with us the next day. Five of those on our list are strongly associated with deciduous trees such as the oak: light emerald; common emerald; iron prominent; the poplar hawk-moth and, a scientific celebrity, the peppered moth...

The peppered moth game

The peppered moth has been much studied. Theirs is a well-known story, so I'll just touch on it briefly here, really by way of an introduction to a lovely game you can play based on this moth. Back in the 1950s, a scientist called Dr Bernard Kettlewell was interested in the rising numbers of the dark versions of the peppered moth, which were first noticed back in Victorian times. He wondered if the appearance of the moths was changing over time in response to their changing environment – in this case, caused by heavy, sooty industry. Kettlewell came up with a hypothesis: that forests near trees that were blackened by soot from factories would house mainly dark peppered moths.

Dr Kettlewell enlisted amateur entomologists across the country to test this idea, and together they showed that, if the moth's colour matched its environment, it had a better chance of survival, thus adding further weight to theories of natural selection. So it was that our humble peppered moth made its contribution to human knowledge. It was a great moment for citizen science.

We now know of other species of moths that have darkened over time in polluted areas. Meanwhile, the peppered moth has been lightening up again (or reverting to the lighter form) as we have reduced pollution from heavy industry.[8]

So, about that game. You can play online. It features our peppered moth and offers a simple and satisfying way to see natural selection working for yourself. In it, you are a bird and your aim is to catch as many moths as you can. Some moths are darker than others. Some moths are flying in front of trees, some with light-coloured bark and others with dark bark. As you'd imagine, birdy-you will find it easiest to catch the moths that stand out – those that are a different shade to the trees. The other moths – those that are better camouflaged – tend to live another day, and if they are lucky, will go on to have offspring. You'll find the Peppered Moth Game on the Ask A Biologist website.[9]

WIDER WORLD

We have already seen how creatures such as birds help keep the nature around the oak in balance. In turn, caterpillars of moths and butterflies play a vital role in feeding birds. As with so many things in nature, it all links up.

Take moths, butterflies and the blue tit. The number of caterpillars the little blue tit needs is astonishing. On BBC Radio 4's *More or Less* programme, the presenting team showed that the UK's young blue tits gobble up an amazing 35 billion caterpillars every year. The RSPB's Dr Malcolm Burgess said the hard-working adults 'feed their chicks at rates of about 35 to 75 visits per hour, so that's pretty much a visit per minute'. This extraordinary number will include one to three caterpillars per feed. And that doesn't even make up the whole diet – spiders, for example, play an important role in the nutrition of a young blue tit.[10]

We can give them a hand, by the way; people who create wildlife-friendly spaces, for example by growing plants with seeds and berries, are helping adult birds to keep going while they forage the live food for their families on nearby oaks and other trees. In our own human way, by enjoying birds in our garden, park or green space, we are helping to keep the wider world of the oak tree going, as long as we keep things clean and healthy.[11]

Meanwhile the humble aphid plays a vital role in feeding caterpillars, such as that of the purple hairstreak, in providing the glorious-sounding food, honeydew. I think there is something sublime in the way honeydew sounds just heavenly and turns out to be aphid poo.[12] Aphids known for their oaky associations include the scarce variegated oak aphid and the blue-green oak aphid. Meanwhile the glorious beastie that is the pale giant oak aphid lived undetected until very recently. It was only in 2012 that the aphid was spotted by naturalist Julian Hodgson in Monks Wood, an ancient woodland near Huntingdon, Cambridgeshire. This creature went unnoticed by humans – apparently for thousands of years – on oak trees, partly due to being hidden by its friend, the brown tree ant. I say friend, you might also say keeper, or even farmer. Matt Shardlow, chief executive of Buglife (formerly the Invertebrate Conservation Trust), explained in the *Guardian* newspaper that the ants are milking the aphids for honeydew, 'moving them from high to low pastures and building shelters for them when there's not enough protection'. This is one of the reasons they are hard to see. The article's author, Patrick Barkham, writes that if aphids are disturbed 'at one of their tree-trunk shelters', the ants move them, 'carrying the smallest individuals in their jaws and hustling the larger aphids down to the ants' underground shelters... the ants keep the aphids underground during severe weather. In summer, when the sap rises, the ants march the aphids up the tree trunk to ensure they are well-fed and can provide the ants with sweet honeydew.'[13]

Again, doesn't this make you wonder about all the other creatures we haven't yet noticed?

···Moth and Butterfly Hunting Tips···

PURPLE HAIRSTREAKS

To see a purple hairstreak, contact your local Wildlife Trust, natural history society or Butterfly Conservation local group. Or check out their website or social media group to find out if there have been sightings in your area and then follow their guidance. Even better, go out with an expert if you can.

Plan to go out on a balmy afternoon or early evening in summer, July perhaps. They tend to fly high up in the treetops, so binoculars are more or less a must, although you might just be lucky enough to find a low oak – one in a hedgerow perhaps – which means you can get reasonably close.

OTHER BUTTERFLIES TO LOOK OUT FOR

The purple hairstreak is one of the most challenging butterflies to find, so if you are starting out, it might be an idea to look for something more accessible. The great thing about butterflies generally is that most of us can easily see some species, and – helpfully – during the daytime too. There are few summer experiences more pleasant than sitting in sunny spot near some flowers in a nature reserve, park or garden and watching them flutter about. It is a lovely way of zooming in on nature; enjoying the worlds of other creatures for a mindful moment. And you don't have to do this alone. A fantastic way to get going is to take part in the Big Butterfly Count, which takes place late July to early August each year and is not only gorgeously meditative but also means you'll be contributing to science.[14] You can brush up on your butterfly identification skills at the same time with its free smartphone app and downloadable chart.

As there are 59 butterfly species in the UK, it is not unusual for people to set themselves the challenge of seeing every one of them. That feels achievable, doesn't it?

MOTHS

In my view moth spotting is harder, but arguably more exciting, because it is largely – although not exclusively – a night-time activity. Also there are so very many species of moths.

If you want to try it at home there is lots of online advice on building your own moth trap. Enthused, I decided to have a go. How hard could it be? Moths are always hanging about lights at night, aren't they? So I confected my own moth trap from bits of recycled things hanging about in our shed: an old, leaky bucket, cycle light, random plastic and the all-important egg-box furniture. I was very excited. I set it all up, shooed away Pepe the pooch – who I suspect might not be beyond the odd munch on a moth – and then popped back later to see the results of my handiwork.

Nothing, nada, not one nibble. My moth trap was completely rubbish. Looking back, I had taken all sorts of wrong turns, including the bike light – moths, it turns out, can be quite fussy about the kind of light they will fly towards. And thank you to Marion for setting me straight on this: ultraviolet light is key.

If you are more successful than me with making or buying your own moth trap, it is best to avoid trapping too frequently so our moth friends' lives aren't too disrupted. And think of your neighbours: while moths love intense lights, those you live close to might be less impressed with a super-bright light at midnight.

A simpler way of getting started comes in the form of a tip from Lancashire Wildlife Trust, who suggest that you recruit friends or family and head out with a torch at night to check your windows, walls and plants for moths.

And another top tip I have gleaned is to start by focusing on a group of moths called The Prominents, which, as well as sounding like a pop group, are quite literally prominent. They are stout, hairy moths with a lumpy look about them and a distinctive dorsal hump – so are a good way into the moth world.

But my best tip of all for beginners is to find some experts to help get you started. If you check out Butterfly Conservation's website[15] or follow Moth Night on social media, you'll see they post information about local events in summer. Of course, check credentials and be very wary indeed about meeting strangers in a woodland, especially after dark.

· · · · · ·

CROWNS TO CROWNS: OAKS AND MONARCHY

If the purple hairstreak is the jewel in the crown of the oak, the oak itself is often seen as king of the woods, and occasionally as queen. There are many oak trees that have been named by local people as Majesty or The Monarch. When it comes to the oak, those royal-sounding words pour out and trip off the tongue. Oaks come with more regal descriptions than you can shake a twig at, and their claim to the throne is fairly undisputed.

In return, the majestic power of the oak has not been lost on royalty or the state. In the eighteenth century, the landscape designer William Boutcher praised the oak as 'the monarch of the wood, the boast and bulwark of the British nation'.[16]

British (by which I include, for simplicity here, monarchs predating the Union of the Crowns in the seventeenth century) queens and kings have long had a penchant for oaks, whether harnessing their symbolic value to boost monarchy and state, sheltering beneath them, hunting around them or – indeed – getting them in the ground.

In *The Glorious Life of the Oak*, John Lewis-Stempel says, 'By associating with the oak, monarchs hoped the tree's majestic qualities would rub off on them.' He describes how royal business was conducted under oaks and that 'Edward the Confessor took an oath under a large oak in Highgate, London, to keep and defend the laws of England'.[17] The Parliament Oak, which still lives in Sherwood Forest, was famously used as a royal shelter, including during a deeply grisly moment when King John paused a hunt in 1212 to convene a meeting to deal with a Welsh rising. The meeting resulted in a ghastly decision to hang 28 Welsh hostages, all boys aged between 12 and 14 years old.[18]

These days, thankfully, the royal links tend to be about celebration. The Windsor family are great planters of trees. Special occasions and commemorations often involve popping a tree, often an oak, into a hole in the ground. Queen Elizabeth II planted 1,500 trees herself. There are pictures on the Internet of her doing so, first as a girl; then as a young woman in a sensible-looking 1940s suit; then again in the 1950s, glamorously teaming shovel with slingbacks. There is photo after photo, a majestic fashion parade ranging through endless matching twinsets, headscarves and rainwear, showing us how to get a tree in the ground in monarchical style.

It was the first Queen Elizabeth who really got things going on the royal oak planting front, though, as John Lewis-Stempel writes: 'In 1580 Queen Elizabeth, acutely aware of the need for quality timber for the navy, assented to Lord Burghley's order to "empale" 1 acre of Cranbourne Walk, in Windsor Great Park, and sow it with acorns – the first record of a deliberate oak plantation.'[19]

And this mutual understanding between majesties and majestic trees endures; our current king, Charles III, clearly loves trees, and I hope and believe his influence will help spread the news of their importance.

Meanwhile, history has given us a number of royal stories involving oak trees; some adventurous, some boozy, some raucous, some surprising, some spooky, some multi-stranded. There are enough of them to fill a whole other book. I'll share a few of my favourites here. For more stories of amazing oaks, royal and beyond, I recommend the books *Britain's Tree Story* by Julian Hight and *The British Oak* by Archie Miles.

The Royal Oak: How a tree saved the monarchy
One of the most famous of the regal oak adventures concerns the time royalty was saved by a big, beautiful oak tree. This tree hero was The Boscobel Oak or The Royal Oak.

This happened during the time of civil war in England back in the 1600s. After King Charles I was executed, his son – young Charles – famously wanted to bring back the monarchy and take back the throne for himself. This was no easy task when your enemies were led by the fearsome Oliver Cromwell, and in 1651 the young Charles took a bit of a drubbing at the Battle of Worcester. Charles fled, but found his escape route blocked. He needed somewhere to hide, hide fast, and sought refuge at Boscobel House in Shropshire. Enter the fabulously named William Careless, one of Charles's soldiers, who came up with the ruse of holing up in an oak tree.

They climbed into the huge tree and sat there regally, probably not unlike purple hairstreaks at rest, even as their foes passed by, searching for Charles and other refugees from the battle. The king, we are told, rested his head on Careless's lap. With the help of the oak they escaped capture and, after a prolonged detour in France, and adventures that, among other things, involved dressing up as a woman, there was a happy end to the story... if you are a royalist. Charles survived and, eventually, regained his throne. And so it was that an oak tree helped lead to the restoration of the monarchy in May 1660.

Careless, the careful king cushioner, earned himself a name upgrade. The National Portrait Gallery, which houses a portrayal of his oaky adventure in an oil painting by Isaac Fuller, tells us: 'Charles conferred distinction on him by changing his name to Carlos and granting him a coat of arms featuring a green oak on the shield. At the restoration he was awarded the proceeds of a levy on all hay and straw brought into London and Westminster.'[20] Fuller's painting was one of many artworks inspired by the episode, which is celebrated not only in paintings but also in ceramics. I have even spied it on some wallpaper. Meanwhile the adventure was captured in verse:

> How from Worcester fight by a good hap, Our Royall King made an escape;
> How he dis-rob'd himself of things that precious were,
> And with a knife cut off his curled hair;
> How a hollow Oak his palace was as then, And how King Charles became a serving-man
> FROM THE BALLAD 'THE ROYALL OAK OR THE
> WONDERFULL TRAVELLS, MIRACULOUS ESCAPES,
> STRANGE ACCIDENTS OF HIS SACRED MAJESTY
> KING CHARLES THE SECOND'[21]

Royal Oak pubs

The other oaky offshoot to the story above is the sprouting of pubs called the Royal Oak. They numbered 421 at a recent count, which means that, in pub name numbers, they come behind only the Red Lion and the Crown.[22] So, while sadly the original Royal Oak tree has died, its name lives on in the hearts and minds of drinkers across the land. Meanwhile this special tree has left other, living legacies; English Heritage says of the land around Boscobel House: 'Today, the lost oak pasture has been restored including trees propagated from the original Royal Oak', and invites people to come and see a descendant planted by the then Prince of Wales in 2001.[23] While I was writing this book the Prince of Wales became Charles III, which I expect bodes well for oak planting. King Charles the Three loves a tree (more of which later in the book).

Oak Apple Day
All those Royal Oak pubs apparently weren't going to be enough to satisfy the national need to have a right royal knees-up. No, we needed more revelry. We, the nation, needed an extra-special day on which to get especially drunk to remember a past monarch hiding out in a big tree. A day of merrymaking was declared across the land and – joy to cecidologists out there, happy gall hunters, all – it involved galls. Hurrah! The big day, 29 May, was called Royal Oak Day or Oak Apple Day.

Relatedly, the famous diarist, Samuel Pepys was moved to write, back on 1 June 1660, 'At night Mr. Cooke comes from London with letters, leaving all things there very gallant and joyful. And brought us word that the Parliament had ordered the 29th of May, the King's birthday, to be for ever kept as a day of thanksgiving for our redemption from tyranny, and the King's return to his Government, he entering London that day.'[24]

I suspect what Mr Pepys is really saying here is, 'Hey, it's party time and we're all invited (as long as you aren't one of those earlier parliamentarian types, obvs).'

Oak Apple Day was a day when people danced. They danced wearing oak leaves and they danced with oak branches. The branches were actually quite a big deal and people got competitive about them. The best branches had oak apples on, and those with the most galls were the best of all. People who carried them were the stars of the party. They were the in-crowd of the moment.

Then there were the uncool kids, the people who didn't do the right thing. They not only lacked great oaken branches bearing gorgeous galls; they didn't even turn up with so much as a caterpillar-chewed oak sprig in their hats. I don't know why. Maybe they had a kind of secret sympathy for Oliver Cromwell and his suddenly-not-so-fashionable views. Possibly they were poor and didn't have the time to put down their tools from whatever desperately hard manual labour they were doing that day to scrape out a meagre living for their family. Perhaps they forgot to pick up a few leaves on their way home from the forest. Maybe they hadn't received the memo about the dress code. Whatever the reason, they risked ending up as persona non grata. A couple of special punishments were reserved for these people. They were – don't judge me, I'm only reporting this – whipped with nettles or had their bottoms pinched.

These maltreatments were taken up with such enthusiasm that Oak Apple Day also became known as Pinch-Bum-Day and Nettle Day. Another – unsettling – name was Shick Shack Day. 'Shick shack' was an insulting derogatory term for nonconformists and in this case referred to people without their oak leaves.[25] Perhaps the most disturbing of all the terms was Maid's Ruin Day.

Owain Glyndwr (Owen Glyn Dŵr/Owen of the Glen of Dee Water/Owen Glendower)

Welsh prince Owain Glyndwr, the last native Welsh Prince of Wales, gave his name – which comes in a few spelling options – to an oak tree, The Owen Glendower

Oak, and a number of pubs too (bit of a sub-theme developing here).

The fact that this particular tree was in Shropshire and not Wales was, well a bit awkward, and might well have come as something of a surprise to Owain himself, had he known about it. As Archie Miles points out in his book *The British Oak*, it might also explain the Anglicised version of his name in this context.

This oak story goes that way back in 1403, Owain forged an alliance with a chap called Henry 'Hotspur' Percy (who, when it comes to excellent names, gives Careless a run for his money) against King Henry IV. A battle was brewing, and Shrewsbury was the focus. King Henry got there first, which gave him the advantage of being able to put himself between the armies of Owain and Hotspur. As Owain approached, he is said to have climbed the oak to take stock of the situation, seen that this was not a battle he could win and withdrawn. Poor Hotspur fell in battle.

The problem with this particular oak tale is that it probably isn't true. We perhaps have to take some of the old tales with a pinch of salt. We can still enjoy the stories though. In this case the tree in question appears to have been too far from the battle to give Owain the lie of the land. More awkward still is that Owain seems to have been in Carmarthenshire at the time. Even the tree's name is up for discussion, as it was also known as The Shelton Oak, among other things. The oak that appears to be at the centre of the uncertain tales died in the twentieth century. Happily though, in a way, it is one of those trees we can still enjoy because it was recorded by several artists, including the famous Jacob George Strutt, engraver and lover of trees, back in 1830, whose work shows a great old beast dwarfing some nearby cattle. It has branches both dead and alive and an enormous bole flowing out to the ground in a wide skirt.[26]

The Bruce Tree
An oak that finds its way into this book mainly because of its quirkiness, The Bruce Tree is sadly fallen now. It was named for King Robert the Bruce and also known as The Strathleven House Oak, and, as Archie Miles says, it is 'just possible to believe that the Strathleven House Oak was growing in the early fourteenth century and hence might have been known to the King of the Scots'.[27] The story goes that King Robert's face somehow found its way into the heart of the tree; in *Britain's Tree Story* Julian Hight recounts how the Strathleven Artizans, a voluntary group that runs the King Robert the Bruce Heritage Centre, had a spooky, face-to-face encounter with a long-dead monarch: while chopping up the tree after it fell because of a fire in 2004, 'the Artizans were surprised to see the image of a face appear within the heart of the great Oak. This cemented their view that it was indeed the Bruce Tree, symbolically displaying a likeness of the heroic Scottish king.'[28]

Elizabeth I
Queen Elizabeth I liked a tree, and there are several oak stories linked to her life. (It's no coincidence that this is the longest section in this subsection.)

It is said she was sitting beneath an oak tree eating an apple in Hatfield Park, Hertfordshire, when she heard that her half-sister, Queen Mary Tudor, possibly better known as Bloody Mary, had died. Loss of a sibling is of course momentous news for anyone, but in the case of this particular 25-year-old the news was especially life-changing, as it meant Elizabeth was now to be crowned queen. I say eating an apple; other reports had her reading a bible. I lean towards the apple story; I can't help but think that the bible version sniffs of medieval PR spin, but who knows?

The oak tree in question died and its remains were removed on 17 November 1978, which was, as Julian Hight notes, 'exactly 420 years after Elizabeth I's coronation day.' Queen Elizabeth II planted a new oak in its place in 1985.[29]

Marvellously, we do have some surviving trees that met the Good Queen Bess. One of them – aptly called The Queen Elizabeth Oak – lives in Cowdray Park, Midhurst, West Sussex, where Queen Elizabeth I once had a picnic and also, apparently, a quarrel, while sheltering during a storm back in 1591. Torrens Trotter writes in *Cowdray: Its Early History*, 'Tradition has it that Lady Kildare (the sister of Lord Montague) incurred the displeasure of the Queen by daring to shoot with her... we read Elizabeth was so annoyed that she (Lady Kildare) did not afterwards dine at the royal table.'[30]

Maybe this was historical gossip. How can we really know? We do know that Elizabeth I liked a good hunt, so maybe there is something in it. Oliver Rackham in his book *Woodlands* says, 'Elizabeth liked to be thought of as "a weak and feeble woman" who could not keep up with her father's manly activities, but in practice seems to have outdone him as the greatest sporting sovereign before George V.'[31]

The Queen Elizabeth Oak can still be seen today and she is a stunner – an ancient sessile who could be as old as 1,000 years, with a great, broad girth and a low-slung canopy. She has the delicious proportions of a plump, portobello mushroom. She squats grandly on a hillside, with a good crown of leaves in summer, and surveys the rest of Cowdray Park. She has a bit of company, an attendant of sorts, as we might expect of a monarch. A little further up the hillside is The Lady-in-Waiting Oak. She is another ancient sessile beauty in the same mould as her queen, possibly a few years younger (we are talking oak time of course, so perhaps 100 years younger), although she is in slightly less good condition so she might not outlive her arboreal monarch. She is hollowed out to the point you can see straight through the middle of her and the rotted wood spews out from her insides in a trail of fine, moist crumbs. She brings to mind a noblewoman who has eaten some venison that disagreed with her.

Another of Elizabeth I's oak friends, The Crouch Oak, resides on a small patch of grass next to a rubbish bin in suburban Surrey. Some say the tree got its name because Elizabeth was caught short during a hunting trip and needed 'a quick crouch'. This may be a mischievous interpretation. Others say the oak witnessed a hunting drama when the queen was charged by an angry stag and saved by a huntsman who gave his life to protect hers. This risks going a bit Herne-the-Hunter. Nowadays this feels a surprising place to find such an incredible plant; however, it is far from alone in being an urban ancient oak. In this case the area used to be part of Windsor Great Park – right at the boundary – which would explain in part how the oak reached its great age, presumably having long lived in a setting where it got plenty of space and light. There is something startling but also lovely about this incredible tree, serenaded as it is on a summer day by cars pumping out music as they pass by, oblivious to her existence. I understand it is much loved by locals – and who wouldn't enjoy having tree royalty on their doorstep?

Many more oaks can claim a link to Queen Elizabeth I. There are too many to mention here, so I'll end with a few words about a couple of notable trees. Elizabeth is also said to have planted The Panshanger Oak, near Hertford, another knockout of a tree. This beauty is not at all in the mould of The Queen Elizabeth Oak, being as different as an aged oak can reasonably be. She is a tall, lengthy, upright beast: as a 'maiden' tree she was never pollarded, which is unusual for a really old oak.

GENDER OAKQUALITY

Oaks are often described as kings, but they can just as easily be queens. One of the many things I learned on my oak journey is that oak trees are arguably both male and female. They certainly have both male and female flowers. The male flowers dangle down in slender yellow-green catkins. The female flowers are tiny, red or brownish and bud-like, exquisite if you care to look carefully.

What would the oak say about its own gender? We don't know, of course. However, I like to think that, should you, could you, ask a big, brawny oak about this topic it would roar a thundering, throaty laugh – the kind of laugh that would rumble all around you and under your feet and ripple through your belly – and then ask you a question in return: 'What is it with you humans and your gender fixations? I am an oak. I am what I am.'

4

Life, Death and Beetling About

THE CREATOR, IF HE EXISTS, HAS AN INORDINATE FONDNESS FOR BEETLES.
ATTRIBUTED TO J. B. S. HALDANE

SHERWOOD SUMMER

It is the cusp of June and July 2021 on a warm morning in a wooded part of Sherwood Forest. The oaks around us have changed radically since the spring. Well, those that are living anyway. Their canopies are a deep, rich green and they are big and full. You could hide a whole king and a small retinue up there, along with a host of other species.

On the ground lies an oak apple, wizened like a shrunken head. Its role as a party accessory is long forgotten.

The birdsong is more languid, softer than earlier in the year. Our feathery friends are no longer in their first full frenzy of flirtatiousness. Their parental duties are sapping at their energy and their music is muffled further by the rich fullness of the natural world in early summer.

The foliage has changed too. At only a few months of being, much of it is showing signs of age. It is working in leaf time rather than the much bigger, much, much longer tree time. Many leaves are getting tough, almost leathery. Some are misshapen, mottled with splodgy spots, yellow with a touch of rust red. Turn them over and you find they are splattered with silk buttons, the galls that look like tiny Cheerios. The old leaves contrast glaringly with bursts of fresh sprigs, new growth. These are fresh and fragile as the ears of newborn babies.

Today I am with the Woodland Trust's Louise Hackett, who is a specialist in conservation on a landscape scale – in my layperson's terms, that means looking at the big picture; at the way we can, or should, manage land and link up our natural world. We are off to meet The Medusa Oak. I have heard this tree is a great character and that she has living treasure hidden in her bark and her boughs.

THE MEDUSA OAK

As you might suspect, The Medusa Oak is named – probably by local foresters – after the female gorgon of Greek myth, she with the hairdo made of snakes. She is the oddest-shaped oak tree I have ever had the joy to clap my eyes on, the quirkiest of quercuses. She has a massive bole, a great old trunk so thick that to hug her would take several people. That's not unusual in an old oak tree of course, but here's where she's different: she has been truncated, literally truncated, cut oddly short at about the waist height of an adult. Her remaining bole flares out towards the ground, forming a skirt made of forest whose bark flows downwards in deep, vertical folds. And then, up from her flattened top, sprout a handful of branches that look like long, slender serpents. Skywards they writhe, winding and twisting into the blue. Someone, at some point, gave our Medusa a brutal haircut. She probably wasn't meant to survive. But survive she has, in her unique way. I have heard a theory that she was chopped down during the Second World War to widen a path. Whatever happened, she was treated ruthlessly, much like her mythological namesake, and, as a result, she too has turned into a kind of beautiful monster.

Louise and I take a turn around the tree, feeling the rough bark and peering into the endless crevices. Among the treasures The Medusa Oak holds is a lot of dead and decaying wood. Within that death and decay you can find life, a lot of life. The term 'dead wood' is, of course, often used to suggest something is a bit useless. It might offer a damning description for, say, an annoying colleague who doesn't pull their weight. But in the natural world, it turns out that dead wood is anything but useless. It is life-enhancing stuff for many species. Dead stuff, is, in short, dead important.

You would be forgiven for thinking that dead wood is dead wood is dead wood. But no, nature isn't so simple. There is more to it than you might think. This oak venture is showing me there are many kinds of dead, and as Louise points out the dead wood's features, shows me its secrets, she teaches me a lot about the less-than-living.

I learn that some creatures like dead branches, while others like shattered branch stumps. Some like a bit of fractured bark. Some delight in those damp pockets in a tree where mini pools form; some dawdle in sap runs where a bit of damage makes the tree look like it is bleeding; others hang out in the rot holes left where a limb has fallen. Some of them are partial to the bits that have fallen off trees; some adore wood that is decaying on the ground; others inhabit the spectral dead tree that is still standing, enigmatically, like a ghost in the forest because trees rot in different ways, depending, for example, on whether they are standing upright or lying on the ground. Nature loves all versions of the rotting process. And then there is location. Ideally you will have dead wood in sunny positions and dead wood in shady positions. Sunny, shady, soggy, dry, mildly damp, upright or lounging on the ground – these all provide different environments for different species and different stages of life.

'For richness of species it is the combination of living and dead wood tissue in a tree that is the ultimate,' says Louise. Ancient and veteran trees have dead wood alongside the living wood, providing a great combination of habitat and food. This is one of the reasons they are so important. Louise explains the changes that take place in a tree's structure: 'As a tree grows, sap wood turns into heart wood and that forms the centre of the tree. In terms of invertebrate assemblages that need dead wood, it is that combination of sap wood and heart wood that creates the peak in terms of species. Then you've got the habitat of dead wood with water, nutrients and sugars available too.'

All of this means that, while we should embrace our *dead* trees, we need to keep our *very old* trees alive as long as possible. When you consider how many species can live in and around an old oak, it makes sense that you need lots of different environments and circumstances – different homes to accommodate the different worlds of so many animals, plants and fungi. Louise says, 'As with any species, diversity and richness come with time.' In the case of the oak it not only has a long life but also a lingering afterlife – as the great Oliver Rackham wrote in his book *Woodlands*, 'A dead oak takes at least a century to disappear completely.'[1] This is one of the reasons it is a habitat extraordinaire.

DEAD WOOD IS DEAD GOOD

Louise and I are admiring the pleat-like folds of The Medusa Oak's skirt when Louise says, 'Ah, look at that!' She is pointing to something that looks to me like nothing at all. Then my eyes adjust as I squint, zoning in on a tiny, shiny, black thing about two millimetres long, sitting in a crease of spongy, decaying matter. 'That looks like a saproxylic beetle,' says Louise, peering hard at it, then looking up at me and smiling.

Saproxylic beetles are special beetles. These are beetles that send people who love beetles misty-eyed and off in search of ancient oaks. I hadn't even heard the word 'saproxylic' before this oak adventure, but it comes up so often around oak trees now I can hardly believe I hadn't come across it before. So what is it? And how do you even pronounce it? Saproxylic, for the uninitiated, is pronounced Sap Row Zy Lick (where 'Row' rhymes with 'toe' and 'Zy' rhymes with 'fly'); *sapros* means 'rotten' and *xylon* relates to 'wood' so saproxylic is all about a connection to the dead stuff. I should say saproxylic doesn't refer only to beetles; other saproxylic species are available, some flies and fungi, for example.

Saproxylic creatures engage with deadness in a number of different ways. Some of them feed on it – dead or decaying wood is fine dining for many. Others have a less direct kind of connection. They might live in dead wood, say, or they might feed on saproxylic species such as fungi that themselves dine out on the deadness. What they have in common is that they rely on dead wood for at least part of their life cycle. In the UK alone, about 2,000 invertebrate species – including a great many beetles – rely on a constant supply of dead and decaying wood.[2]

Some of the beetles would need to be dissected to be identified – like the shiny living dot in front of us now. I try to imagine how on earth you'd even begin to cut this up. But that's not going to happen. We leave it safe, if unnamed.

BEAUTIFUL BEETLES

Louise and I move on from The Medusa Tree towards a much more open part of Sherwood Forest, Budby Heath. The trees are no longer crowded together but dotted about the landscape.

We sit down by a sandy path to enjoy a snack. The sun is picnic-perfect now: warm but not too warm, just enough to soften the chocolate on our digestive biscuits. Just right. We are settled, at al fresco oneness with the big, wide, wonderful world, when a wild thing decides to visit us.

It is as I open my notepad, its white pages suddenly dazzling in the sunshine, that our new friend appears, plopping onto the paper with a satisfying *pok*. It is a summery kind of sound, like the buzzing of bees, the faint fizzing of a glass of lemonade or the note of a distant cricket bat sweetly meeting its ball.

And hurrah! We can identify this beetle friend with the help of the book I'm carrying: *Britain's Insects* by Paul D. Brock. Even more helpful: it has a name I can pronounce. It is the spotted longhorn (*Rutpela maculata*). It is a saproxylic beetle. And, for me, this little character goes on to open up a new, beetling world, one that involves a journey through the shadow of death and beyond.

The spotted longhorn beetle – also called the black-and-yellow longhorn – is, as its names suggest, bright and colourful. And it is a long creature, all legs and endless antennae. If beetles had supermodels, this would surely be one. Its chic outfit of spots and stripes and its fine hornlike antennae, stretched out in front – are all panache, banded in their black and yellow as they are.

This is an insect that could do with a catwalk and, in a way, it has one. This comes courtesy of the umbellifers, that set of plants whose flowers form great lacy masses of blooms, species such as hogweed and cow parsley, also known as Queen Anne's Lace for its frilly, spilly extravagance. Great plate-sized lacy landing pads are a perfect stage from which to show off a snazzy beetle outfit. I wonder if our little longhorn friend mistook my startlingly white notepad for just such a flower head.

A useful thing I learn about UK longhorns is that they are a good starting place to get into beetling – the adults are fairly easy to spot, reasonably easy to identify with a good guidebook, and it is a joy looking for them in late spring and summer, or even not looking for them but just sitting in the right place, hands sticky with melting biscuit or perhaps busily stroking a sun-warmed pooch such as Pepe. I learn from my book that while the adult longhorns like flowers for their sweet nectar, in many species their young, the larvae, love rotting tree stumps. That is where the eggs are laid, where they hatch and then feed. They spend two or three years down in the damp, dank rotting wood of oaks – as well as other broadleaved trees including birch – before they grow up and seek out the wider world, find a mate and breed. Their adulthood is brief – it lasts a few weeks – but it is beautiful.[3]

We are rich in weeds where I live. There are fields and hedgerows abundant with daisies and brambles and plants like hogweed. One person's weed is another's joy, right? And so it goes for insects, and this all makes a great reason for an insect-spotting picnic on a warm day.

Longhorns are 'of great ecological importance in many ecosystems, many species provide an invaluable pollination service and the larvae eat decaying matter, in turn recycling nutrients through the ecosystem,' says the UK Beetle Recording team.[4] The life story of our new friend highlights this in spades. It shows how nature needs both dead and dying bits of trees, and that it also needs the big wide world beyond them, in this case nectar-rich flowers. It offers a study in the need for a joined-up natural environment, one that connects its living parts, dead parts and blooming parts. In nature, nothing is wasted.

This has important implications for the way we manage our natural havens such as woods, trees and their surroundings. I once attended a talk by naturalist Steven Falk, winner of the Royal Entomological Society's Marsh Award for Insect Conservation, who combines his expertise on the natural world with art and photography. He said, 'The great thing about invertebrates is that it makes you think a bit laterally, not just about the site but what's going on around that site.'

Unfortunately not everyone thinks that way, not yet anyway. For example it isn't unusual to see hogweed cut down – 'tidied'– early in the season. Sadly this can deprive a raft of wildlife like these beetles of important nutrients. It is like smashing up our creatures' food cupboards just when they need them most.

· · · Tips for spotting longhorn beetles · · ·

- The important thing about this activity is to find yourself a lovely little sunny spot in which to have a nice sit down with a cup of tea, perhaps some cake, in late spring or summer. Ideally do this by some dead wood, perhaps a log pile, one with broad-leaf logs such as those from oak and birch. If you can find such a spot near nectar-rich hedgerows such as those with hawthorn, hogweed and cow parsley in flower, that would be ideal. Health warning: if you go rummaging among the umbellifers, be sure to avoid giant hogweed (*Heracleum mantegazzianum*), which can cause really awful blistering. This non-native can be confused with other plants such as our native hogweed (*Heracleum sphondylium*), but it can be distinguished by its size and blotchy stem, among other features. There is a lot of information online that can help.[5]
- Have to hand some way of identifying the beetles. The FSC fold-out guides are lightweight and really handy for this.[6]
- Hopefully you will see some beetles that you might want to log in a notebook (a log log perhaps) or, even better, the app iRecord, in which case you'll be contributing to our knowledge of the world, and if you don't, well you are sitting in a sun-warmed spot with a nice cup of tea.

Win-win.

· · · · · ·

CONFESSIONS OF A WILDLIFE GARDENER

We crouch down on the forest floor to inspect a fallen branch. Louise touches some flaking wood with peeling bark, saying, 'This particular food source has been used up here.' This comment makes me do a double take. Like many gardeners, I have created a log pile for the food and lodging it offers to beetles and other creatures. It had never occurred to me that the logs had a shelf life, although the minute this is pointed out to me it feels kind of obvious.

Louise continues, 'People that think they can leave their log piles, job done. As soon as that bark starts peeling off it stops providing that ultimate habitat. It's not that the dead wood doesn't provide benefits, but, if you want to support the broadest range of species, really you need to be renewing it every couple of years.'

Cue gardeners' guilt: I think of the lesser stag beetles (the gorgeously named *Dorcus parallelipipedus*) that we were excited to find at home one year. We first noticed them when a rotten log fell apart, revealing, like hidden gems, their great big, squidgy-looking C-shaped larvae.[7] I thought I was looking after them. I made log piles and a protective ring of packed twigs – a kind of dead hedge – to keep Pepe the pooch at bay, just in case he decided they might make a novel kind of Scooby snack. But I haven't added new logs in years. It occurs to me I haven't seen any lesser stag beetles for a while. Have I – proud wildlife gardener that I am – let their homes just dissolve without offering any new accommodation? What other creatures might I have made homeless? I feel bad. I resolve to replenish my insect homes right away. A bit of online research tells me that while lesser stag beetles like a bit of oak, they also enjoy rotten ash, beech or fruit wood. This reminds me of another good tip: to vary the kinds of wood in insect inns.

I learn I'm not alone in my dead wood neglect. In the 'State of the UK's Woods and Trees' report, the Woodland Trust says just 7% of native woodland in Britain is currently 'in ecological good condition'.[8] Only 7%! Our woods could be so much healthier, and in turn house so much more wildlife. Part of the problem is not having enough dead wood. In part this is about our human habit of tidying up, clearing dead stuff away, not always realising how life giving it is.

RARE BEETLES

Charles Darwin is believed to have said, 'Whenever I hear of the capture of rare beetles I feel like an old war horse at the sound of a trumpet.'

Our spotted longhorn is not a beetle to have Darwin hearing a trumpet. However he might well have felt like a battle-hardened steed answering its call had he found himself in a certain garage in Neath Port Talbot in 2012. There a local resident came across a blue ground beetle (*Carabus intricatus*) in a log pile. Buglife tells us this beastie had never been recorded in Wales before, and in fact it had been thought extinct in the UK until it was rediscovered in Dartmoor in 1994.[9] And what an animal! The blue ground beetle is one stunning beetle. It is big, beautiful and a deep metallic blue or purple. It is Britain's largest ground beetle, and also one of the rarest. Buglife confirmed what it was, and then wildlife surveyors soon also found it to be in nearby Coed Maesmelin, an ancient oak woodland. It seems to like damp oak and beech woodlands with south-facing slopes – probably ancient pasture woodlands, with lots of old trees and lots of dead wood.

Another beetle that might make coleopterists hear trumpets is the Moccas beetle (*Hypebaeus flavipes*). This rare beetle is so closely associated with only one site, Moccas Park, that it bears the park's name. If it was ever to be discovered in the UK outside the park, Darwin might hear a whole orchestra.

WHISPERING TREES

Later in the day with Louise in Sherwood Forest, we find ourselves away from the open spaces of the forest, wandering through dappled light and towards The Major Oak. Louise tells me she loves the way you can read the history of a wood as you walk through it. My quizzical expression prompts an explanation. She gestures towards the great tree itself, The Major Oak, which is now in front of us, 'It is in the ancient and veteran trees themselves. You'd expect to find these trees in a more open environment.'

The old oaks, you see, wouldn't have reached their great age in a more closed woodland; they wouldn't have had the space and light they needed to help them reach that age. Louise tells me old records including photographs help confirm this. Later, I go online and look up pictures of this long-celebrated tree. One dates back to the Victorian era.[10] Ghostly-looking sepia people stand proudly in front of the famous tree. Around it is great space and light.

So this tree is telling us that this spot was once a much more open space, more like parkland or wood pasture, wood pasture being the kind of landscape where trees are dotted around – some young, some old – along with shrubs and scrub. There might have been areas of grassland or heath. Parts would have been grazed by animals.[11] Now that I've learned to read a tree's story, I think this will always stay with me. Later I come across a quote by W. H. Auden: 'The trees encountered on a country stroll / Reveal a lot about a country's soul.'[12]

There is a clue in the word 'forest' too. Louise explains that forests in the old sense, weren't 'wall-to-wall trees', which is how many of us picture them. 'We need to stop thinking of them as just trees,' she says. They were a patchwork of different spaces, places with different characters. Woodland Trust conservation volunteer Eleanor Clark explores the theme in a blog posting titled 'What's the Difference Between a Wood and a Forest?'

> The modern day understanding of the term 'forest' refers to an area of wooded land, but this has not always been the case. The original medieval meaning was similar to a 'preserve', for example land that is legally kept for specific purposes such as royal hunting. So 'forests' were areas large enough to support species such as deer for hunting and they encompassed other habitats such as heaths, open grassland and farmland.
>
> The term woodland is also considered to be land covered with trees and vegetation, but in the UK woods tend to not be as large as forests. For example, Loch Arkaig pine forest in the Highlands of Scotland is 2,500 acres, while St. John's Woods in Devon is just three acres.[13]

Even Robin Hood is making a bit more sense to me now. All the charging about on horseback by Errol Flynn, Kevin Costner and friends couldn't have happened in a tree-crowded wood, could it? As Carl Cornish – whom we met in Chapter 1 – pointed out during another walk in Sherwood Forest, nowadays you wouldn't get far without meeting a low branch.

Louise and I walk away from The Major Oak. In her book *Wilding*, the serendipitously named author Isabella Tree says: 'To supplement natural regeneration, forest officers in the seventeenth century were instructed to "caste acorns and ashe leyes into the straglinge and dispersed bushes; which (as experience proveth) will growe up, sheltered by the bushes, unto suche perfection as shall yelde in times to come good supplie of timber". So important were thorns and holly to the regeneration of trees that a statute established in the New Forest in 1768 imposed three months' forced labour on anyone found guilty of damaging them, starting every month with a number of lashes of the whip.'[14]

So, left to their own devices, woods will shift around the landscape. They will follow the trees and grow up around them. And they will leave ghost woods – the remnants of old woods. You can see ghost woods more clearly in Scotland, where there is a more open landscape. Oliver Rackham in *Woodlands* says, 'Scotland is another country, and even the nature of woodland is not the same as in England. Woods less consistently have sharp edges. Pinewoods and birchwoods have a history of moving about within the same general area, but even the oakwoods were not often sharply demarcated from moorland.'[15]

In a way woods really can walk then, so maybe Shakespeare's Birnam Woods plot twist wasn't actually that far-fetched after all.

Then, referring to Nottingham Castle, which is now surrounded by pubs, houses, office buildings and roads, Louise tells me: 'The historic extent of Sherwood Forest [the original area subject to Forest Law] would have extended right up to the castle wall and even further south of the city. Although the boundaries flexed and changed over time this area is still considered to be Sherwood Forest – it doesn't look as it once did but it would be amazing to have something similar again – although it would have to look very different, as a lot of people have their homes and livelihoods established across the landscape now.' We talk about the New Forest, which encompasses villages and other human habitats.

If you haven't visited the New Forest, I would urge you to go. It has a wonderful blend of trees and woods and open land and homes for people and homes for wildlife, and you even get big animals – ponies and cattle – wandering around nonchalantly, as though being in this together is just no big deal. It is a joy, and it is this matrix of habitats and life that Louise is getting at.

So why not indeed link the city of Nottingham with Sherwood Forest? The Woodland Trust and partners are already working to link up the landscape in 'Treescapes' such as the burgeoning Northern Forest, which will stretch from east to west across the whole of northern England, from Liverpool to Hull, like a belt that takes in Leeds, Manchester and Sheffield. Similarly, The Wildlife Trusts set out a vision of Living Landscapes.[16] No one is suggesting we remove towns and cities. Instead the idea is that we work with them, in them, around them, to link up the landscape with natural space so that it works better for wildlife and better for people.

To turn this kind of vision into reality you need to be sensitive to local conditions and local communities. 'Every forest is different from the others in the way every person is different and every tree is different from the others,' says Louise. Geology, she explains, is key to the character of a forest, in fact it is often the key to why a forest is there in the first place, or rather why it remained in the landscape when you take into account human development over the last few hundred years. Often forests are in the places that weren't particularly good for growing food. Sherwood, for example, has very sandy soil. In the past there would have been much less agriculture in the landscape around what remains of the forest. But this changed in recent years; with the advent of modern agriculture, fertilisers have made this a rich farming area.

More widely, the health of our forests links to the general health of Britain and, bigger still, the well-being of our planet. Whoah, you might say, what about farmers? Why would a farmer give up some of their productive land to create new habitats unless there is an incentive to do so – they need to earn a living and we need to eat. The fact is we need a landscape that can support both nature and agriculture. This is where we need great decision making at government level – local, regional and national – because landowners need support if they are to work with nature for the benefit of everyone.

WHAT HAVE THE NORMANS EVER DONE FOR US?

Well, for one thing, they played a role in helping some of our saproxylic beetles, albeit inadvertently. Horrendous as it undoubtedly was to be attacked back in 1066, this pivotal date in British history when the Normans invaded led to a set of conditions that helped some of our oaks to live to a great old age. And as old, old oaks have all kinds of dead and dying parts, they in turn are vital for some of our saproxylic species, including our beetle friends.

The story – in brief and in my very simplified terms – goes like this. The Normans created many deer parks, particularly in England. Many of the trees in the parks were dotted through the landscape, rather than being hemmed into denser, darker woods. As we have seen, oaks like space and light. Some of these trees thrived and survived, and we think the odd one or two alive today – such as The Major Oak – might even have been alive as young trees way back at the moment of the Norman invasion in 1066, nearly a thousand years ago. King Offa's Oak and The Parliament Oak might be even older, although there is, of course, much debate about their ages.

Of course those Normans almost certainly weren't thinking about us or the ancient trees of the future at the time. They were probably more interested in their stomachs and a good meal. They loved a bit of venison, and we have seen how important deer were to posh, medieval entertaining, as highlighted by tales such of that of Herne the Hunter. A fair number of these deer parks remained largely in the hands of the aristocracy. They tended to have the all-important open areas with trees dotted about. While some have been chopped up into small pieces of land over time and others have been lost along with family fortunes, those that remain tend to have been subjected to a reasonably consistent style of management.

In the special areas where you have ancient and veteran trees, you get a great richness of species, including invertebrates. And, to give all of them a future, we need a supply of trees of different ages nearby, successive generations that will take on their forebears' role of providing new homes with similar conditions. We can tell where this has happened over time because, like oaks, rare beetles can tell us something of the history of the landscape. When you know that some species can only disperse a couple of hundred metres and you find pockets of them dotted about the landscape, you can say with confidence that big veteran trees with heartwood decay were once fairly close to each other in that area – because these creatures need these aged tree 'stepping-stones' in order to travel. That takes us back to the deer parks and areas with similar conditions. I should repeat the health warning that I have vastly simplified this – landscapes are of course very complex things – but well, here's something the Normans did for us. Santé to them.

ANCIENT OAKS AND FUTURE ANCIENT OAKS

Excitingly, research suggests there are many really special, ageing oaks still to be discovered and, within them, the multitude of wildlife they host. Dr Victoria Nolan of the University of Nottingham and colleagues have found that, while the Ancient Tree Inventory (ATI) has an impressive total of over 190,000 records, it may document less than 10% of the nation's oldest trees.[17] This is astonishing to me. There could be two million ancient and veteran trees in the UK, and most of them are yet to be found. Cue a tree treasure hunt. We can all get involved in finding and mapping these special trees, all be tree discoverers. If you'd like to join in, check out the Ancient Tree Inventory online.[18]

Mapping these tree treasures is vital. If we don't know where they are, how can we protect them? We have seen how important it is to keep our old trees alive for as long as possible, and yet they slip through our fingers: 73 ancient and 393 veteran trees were marked as 'lost' on the Ancient Tree Inventory between 2010 and 2020. These are just the trees we know about – those that are reported. Each time we lose an ancient tree we risk losing the species it houses. The Buttington Oak once stood on Offa's Dyke in Wales and collapsed in 2018,[19] after which a fungus new to science was discovered on it. And what of the veteran and ancient trees of the future? We are losing these too. A recent study in the Eastern Claylands of Suffolk and Essex showed we have lost half of our trees outside woods.[20]

The ancient generation gap
Old trees, dying trees, dead trees; they are all important, as are young trees of course. But what happens when you get a tree generation gap?

If nature was left to do her thing we'd have a reasonably even spread of ages of trees. But our landscape is subject to human-led trends and traumas and this can mean that, when it comes to trees, we end up with missing age groups. This has happened at Sherwood Forest. We have the dead trees and the ancients – trees 400 years old or older. The next generation with considerable numbers is of trees around 200 years old – they result from a push to plant trees around the time of the Napoleonic wars, when oak was in high demand for the ships, those 'wooden castles' as they have been called, that were so important at the time. But as you can see, we have a considerable gap between these two generations. Sherwood Forest now faces a sudden drop-off in the ancients – some of which are near the end of their natural life – and also the dead ancients, which are in the process of decaying and will eventually disappear. This creates a risk for the rare species in the oldest trees. When one tree home disappears they need another tree with similar conditions, and yet the next generation may not be old enough – with all the different conditions an ancient tree provides – to cater for all those needs.

Future ancient trees
One way of helping nature is an emerging technique called veteranisation – in essence, speeding up nature's job.

What happens is that younger trees, where abundant, are deliberately 'aged' in a sense, to provide the habitats some species depend on. It is about creating features such as holes where fungi can take hold, and where birds, bats and other creatures can find places to nest and take refuge. Veteranisation takes different forms. It can be as simple as carving a cavity into a tree, a home for a bird or a bat – human woodpeckering, if you will. Or it can take a more dramatic form such as a 'lightning strike', where you take a slice out of the tree in a downward sweep. It might involve 'ring barking' a branch: removing a band of bark all the way around, which means the branch will die and rot. It sounds brutal, but its advocates recommend its use only in places where those younger trees would be removed anyway, say in crowded areas where you need to bring in more light. Veteranisation should only be undertaken by experts and never on existing veteran or ancient trees. Mother nature does it best, and should be left to her own devices where already working her magic.

A trial to evaluate the effects of veteranisation began in 2012 and so far, the results are encouraging, with birds and bats flocking to these new des-res prefabs. In trials in Sweden, England and Norway, some species responded very quickly; more than 60% of the nest boxes were used in the spring after the holes were made. Woodpeckers, bats and bark beetles all seemed enthusiastic to take up residence, and there was a huge hike in the all-important fungi species, so vital in speeding up the decaying process.[21] The work and related studies continue, and its advocates say that while this will never replace our ancient and natural veterans, it is one way of helping in dealing with a challenge and bridging the tree age gap.

GODS OF SMALL THINGS
At the end of our day in Sherwood, Louise and I linger by a big, deceased oak tree. It is tall, grey, and appears to be sculpted by nature itself. Louise and I peer at it hard, at its great cracks and crevices, at the tiny holes in its wood. Some of the holes are like pin pricks, others are a few millimetres across. We know these are caused by boring insects.

When I say these are boring I mean, as James Canton has fun pointing out in his beautiful book *The Oak Papers*, we are talking about wood-boring, not, you know, dull-boring. And for the woodworm, life is full of drama and jeopardy. I should mention at this point that the woodworm isn't really a worm at all but rather the larva of a beetle, perhaps an oak pinhole borer.[22] Anyway, imagine being that little, badly named creature. You are burrowing away inside a cavernous old tree, with its endless tunnels and dark chambers like a tiny, invertebrate Indiana Jones, and danger could be around any corner, possibly in the form of a bigger beetle.

We only know such things because of the amazing people who sweat the small stuff, the hidden stuff. I'm talking about the people who explore these worlds. Their astonishing diligence is both admirable and surprising. They draw attention to creatures that are often almost invisible, but vital to our big world. They open up our universe. These people, the entomologists, cecidologists, coleopterists and beyond, to me, are, to take inspiration from a book title by the great writer Arundhati Roy, 'gods of small things'.

These people are my heroes. They seek out the teeny, tiny, obscure creatures that aren't necessarily the things that grab our attention, because, well, they are teeny, tiny and obscure. When I say 'obscure', I mean obscure to we humans. They will probably be much more visible to, say, a blue tit or a nuthatch. These people get to know the little characters that aren't easy to know, while the rest of us barely notice them. And they do battle with the Latin or Greek names that are hard to remember (for many of us) – many of these tiny creatures haven't been given the courtesy of a nice, descriptive, easy-to-remember name, not yet anyway.

We still have so much to learn about our trees and woods and how they interact with and influence the wider environment. We continue to discover new species in and around them. We are only beginning to understand their vital role in long-term carbon stores, especially in the soil around and beneath them. So what else are we squandering when we lose the generations of trees that so many tiny invertebrates call home? At this point, we simply don't know. Power to the leaf-littered elbows of the gods of small things – may they be given research grants and the support they need to continue their vital explorations into unknown worlds.

5

Enchanted Forests: Folklore, Myth and Magic

RELICS OF AGES! COULD A MIND, IMBUED
WITH TRUTH FROM HEAVEN, CREATED THING ADORE,
I MIGHT WITH REVERENCE KNEEL, AND WORSHIP THEE.
IT SEEMS IDOLATRY, WITH SOME EXCUSE,
WHEN OUR FOREFATHER DRUIDS IN THEIR OAKS
IMAGINED SANCTITY.

WILLIAM COWPER, FROM 'YARDLEY OAK'

GODLY, KINGLY AND THUNDEROUS
Listen to an old oak tree and you might hear the creaks and groans as it adjusts to the strength of the wind. You might hear the beat of the birds that drum on its branches. You might hear the breezes that find their voice as they blow through mosses, lichens and nests. And if you listen well enough, it is said, you can hear the songs of the fairies.

We read about folklore and fairies in books and we smile at such fanciful ideas and at how quaint it all was back in the day when people believed in such things. Then, one day, you find yourself alone in the heart of a forest, in the midst of nature. And then the breeze breaks through a gap and strokes your face and a twig snaps nearby. And you might start to think about the things, the non-human things, that are responding to their surroundings.

And if the light starts slipping away and the dark inches closer and your heartbeat quickens a little, at such moments those sweet old folk tales and fairytales, the myths and the magic, the legend-blurred histories, seem more believable and perhaps less endearing than they did when you read them by bright, electric lighting in the safe surrounds of a warm home. At that instant I like to think we are a little closer to our ancestors, a little more likely to understand their oak-fuelled beliefs, their hopes and their fears. Is this magic? If you step through the oak doorway will you enter a new world?

Let's journey with the oak beyond the material world, into religion, myth and magic. And then into some of the most magical oaks and woods.

TO KNOW IT IS TO DRUID
We can't not mention the oaks' namesakes (arguably), the Druids, here. And yet... we know so very little about them.

While Druids are sometimes called the 'knowers of the oak', our information on them is as thin as a freshly unfurled leaf. If only they had written everything down. What we wouldn't give for a bit of gall-ink literature from these Celtic religious leaders. So, what do we know? Well, we know that the word 'Druid' comes from *druir*, the Celtic for oak, or the Greek *drus*. And *wid* means 'to know', so 'Druid' seems to equate to oak or tree knowledge. Or does it? Even that is debated, with others translating the term as 'strong see-er'.[1] We believe the Druids used oak groves as places of worship; they made crowns from oak leaves, used their branches for staffs and the wood to carve sacred effigies. And it seems that they liked a bit of oak-hosted mistletoe. In *The Golden Bough: A Study of Magic and Religion*, Sir James George Frazer wrote, 'Among the Celts of Gaul the Druids esteemed nothing more sacred than the mistletoe and the oak on which it grew: they chose groves of oaks for the scene of their solemn service, and they performed none of their rites without oak leaves.'[2] I presume that because it is unusual to see mistletoe growing on oaks, this made it all the more special.

It is intriguing to guess what the knowers of the oak knew. Did they have a mental list of the 2,300 species that live on the oak? Perhaps they knew of more: 3,300 or 4,300? Did they have long-forgotten oaken and fungal cures for terrible illnesses? Did they understand bacteria? Did they know if purple hairstreaks are rare or plentiful high up there in the oaken canopies? What did they know that we have forgotten?

We know others have revered the oak over time, including Norse, Greek and Roman people. The oak has been associated not only with gods and kings but also the kings of the gods. Frazer wrote, 'Both Greeks and Italians associated the tree with their highest god, Zeus or Jupiter, the divinity of the sky, the rain, and the thunder. Perhaps the oldest and certainly one of the most famous sanctuaries in Greece was that of Dodona, where Zeus was revered in the oracular oak. The thunder-storms which are said to rage at Dodona more frequently than anywhere else in Europe, would render the spot a fitting home for the god whose voice was heard alike in the rustling of the oak leaves and in the crash of thunder.'[3]

There are other types of oak-related worship. In Slavic culture, Frazer says, the oak appears to have been the sacred tree of the thunder god Perun: 'It is said that at Novgorod there used to stand an image of Perun in the likeness of a man with a thunder-stone in his hand. A fire of oak wood burned day and night in his honour; and if ever it went out the attendants paid for their negligence with their lives.'[4] Similarly, Perkunas or Perkuns, the Lithuanian god of thunder and lightning, was associated with sacred oaks.

STRANGE SPECIES

Meanwhile, in some cultures oaks played host to spirits, pixies and fairies; folkloric species that didn't make it onto PuRpOsE's list, that catalogue of 2,300 oak-related beings mentioned in the Introduction. Some people say that the fairies moved into oaks when Christianity came to Britain and Ireland. If you choose to visit to seek their help, just look for a fairy door – a hole in the trunk – and give it a rub with your hand and leave a lock of hair as an offering.[5] Some people believed the oak trees themselves were conscious, and might react strongly and quickly to rough treatment by humans. Sir James Frazer writes, 'If trees are animate, they are necessarily sensitive and the cutting of them down becomes a delicate surgical operation, which must be performed with as tender a regard as possible for the feelings of the sufferers, who otherwise may turn and rend the careless or bungling operator.'⁶ On being felled, the oak was sometimes said to give a shriek or a groan that could be heard a mile away. With apparently tender concern, some people wielding the axe recited soothing verses or apologies before they did the deed. On other occasions they just made excuses, apparently. From what I've read, these might go along lines of 'the priests made me do it.'

The oak was sometimes worshipped not for its association with other gods or souls, but for itself, a practice that prompted a famous bit of tree vandalism more than a thousand years ago. The story features Saint Boniface of the Christian religion. Frazer tells us that 'a sacred oak near Geismar, in Hesse, which Boniface cut down in the eighth century went by the name of Jupiter's oak (*robur Jovis*), which in old German would be *Donares eih*, "the oak of Donar".'⁷ Historian Dr Eleanor Janega picks up the tale in BBC Radio 4's *You're Dead to Me*: 'He [Boniface] is hanging out in the German lands in the eighth century and he comes upon a group of pagans who worship a giant oak tree and, in order to stop them from doing that and to convert them to Christianity, he goes up to it and he gives it a whack with an axe and everyone shows up to be, like, "Please don't kill our sacred tree, dude," but then the entire, huge tree falls down and it splits into four bits and everyone is like "Oh it's a miracle, homies, I guess I do believe in God," so basically that's a way of connecting Christianity to trees more generally and a kind of co-opting old pagan things to say, "Well, here, Jesus has come, you can still have your tree but we are Christianising it" and St Boniface is directly linked to that.'⁸

GOSPEL TRUTH
When it comes to Christianity, the oak tree had its practical uses, with 'gospel oaks' serving as venues beneath which the

preachers could hold forth. One of these, now long gone in physical form, lives on in the sense that it gave its name to the North London area of Gospel Oak. It is said to have sheltered preachers including the early Methodist John Wesley, and served as a boundary marker, standing as it did where the parish of Hampstead met that of St Pancras.[9] Archie Miles tells how 'the tree featured in the annual beating of the bounds, when a priest would lead the community on a walk around the parish boundaries'. Like other gospel oaks, it might have previously hosted pagan ceremonies. Miles recounts how oak-sheltered meeting places saw the balance of power shift from Druids to bishops.[10]

DEVILRY, DANGER AND PROTECTION
Some oaks have been accused of housing devils, and others have inspired acts of sorcery. In *Oak Notebook*, Robin Harford tells us tales of oak-related magic: 'A witch trial is recorded of a witch accused with teaching a woman how to kill a cow by breaking an oak twig and placing the pieces upon the animal.'[11]

On the other hand, as Robin points out, the oak has also been invoked as protector against witchcraft and devilry. He tells us that acorns were said to 'bring luck, health and protection'. I have read that acorns can ward off disease; wearing one on a string around your neck would apparently prevent premature ageing. In *Discovering the Folklore of Plants*, Margaret Baker tells us, 'An acorn in a bridegroom's pocket endowed him with long life and the necessary energy for his new obligations.'[12]

In some places, rubbing your left palm on oak bark on Midsummer's Day was said to keep illness away all year.[13] In others, you could scare away fairy folk by drawing a protective circle around yourself with an oak sapling.[14]

A piece of lightning-struck wood came in handy too as a talisman for protection. Lightning pops up again and again in oak world over time, perhaps not surprising when you consider that, in the past, oaks were often the tallest things in the landscape, a target for a lightning hit. And so people sometimes kept oak twigs, acorns, and oak apples to guard against lightning strikes.

Oaken magic even reached its twiggy feelers into the world of sleep. I have read that dreaming about oaks can be a hazardous experience or a happy one, depending on the nature of the dream: climbing an oak meant a loved one would have a hard time; taking a nice rest under an oak tree suggested long life and wealth but a fallen oak signified the loss of your love. On the bright side, as we have seen, a dead oak makes a lovely wildlife habitat.

GHOSTLY OAKS

Some of Britain's famous old oak trees have been associated with ghosts, devilry and the dark arts. The Big Belly Oak in Savernake Forest, Wiltshire, is one such.

It is said that if you dance 12 times anticlockwise around The Big Belly Oak – naked – you will summon the devil. This old saying came back to me one July evening in 2021 as I approached it alone in the fading light (not nude, as it happens). I know, I know, it isn't sensible to be alone in the forest at this time – if venturing out late I'd recommend doing it in company – but I was later than I meant to be and I am rarely in the area, so didn't want to miss the chance to meet a celebrity oak.

So there I am, up to my ears in scrub and bracken. I can't see much beyond a few feet in front of me and I'm inching my way along through the woodland, eyes focused on the ground. Suddenly I see a dark, swollen mass blocking the path ahead. I raise my head and there, looming above me, is a big, bulging, beast of a tree. It is wearing a great rusting girdle of iron, which looks like a medieval chastity belt, around its distended midriff. In this bloated stomach, dark holes gape like contorted mouths, while up aloft, immense, twisted limbs loom above my tiny human form as if The Big Belly Oak means to frighten me.

It is frightening me.

I decide that, on this occasion, I won't strip and dance around the tree 12 times. Interesting as it might be to meet Satan himself, I'll forgo the opportunity. It isn't only the fear of conjuring up the Prince of Darkness, but also the way that the A346 cuts through this forest and is so close to The Big Belly Oak that the old tree is practically kissing the road. If there are more dangerous things to do than dancing with the devil, I'm guessing a starkers samba on a busy road is probably up there.

As I work my way back to my car, stumbling around in the increasing darkness, tripping over roots, the sounds of the road retreat and, not for the first time, I recognize how easy it is to lose your sense of direction in the forest. Something rustles nearby. I jump. It sounds big. I hear a strange cry and then another. I tell myself these must be the sounds of muntjacs, the small, non-native deer that are now commonplace in parts of the UK. Their cries have been compared both to the barks of dogs and to the sound of someone being murdered in a forest. It is time to go home. An Internet search a little later that evening confirms muntjacs are in the area. All the same, I'll take somebody with me next time.

Not to be outdone by plant-related terror, we humans have, over time, invoked the oak as an accomplice in a number of horrific deeds. Some big old trees have taken on gruesome names such as The Strangling Tree and The Hanging Tree. The Clachan Oak, the ancient sessile oak at the entrance to Balfron, Stirlingshire, was said to be the setting of some grisly scenes. This tree is held together by iron hoops. There are tales that, in times gone by, wrongdoers were chained to these hoops, stuck there while people shamed and ridiculed them. It is said that the apparatus used to attach people to the hoops was a metal collar, connected by a chain: all the better for keeping your petty criminal where you wanted them while subjecting them to a spot of unpleasantness. On one particularly gruesome occasion, it is said that a poor, captive woman fell and was strangled by the iron collar. Left there, forgotten, she died.[15]

OAKS IN CLOAKS

Every woodland can be magical. Every tree has its own charm. But for the most enchanting, fairytale-like experience possible, I recommend a visit to an oak-rich rainforest, a Celtic rainforest.

Celtic rainforest. When I first came across the term 'Celtic rainforest' I had to check what I was hearing. I couldn't quite believe we had rainforests in the UK. Yet here they are, our own temperate rainforests. They are as precious and extraordinary as tropical rainforests; they are sprinkled like rare jewels along the western edges of our lands and they are opulent with biodiversity.

The UK's rainforests tend to have clung on in places that are tricky to reach – remote places, ravines, that kind of thing – and in large part they owe their survival to this. Washed with wet, clean air coming off the Atlantic, their trees, their crags, boulders, nooks and crannies create great conditions for certain species, including lichens, mosses and liverworts. Within them the oaks are draped, festooned with other species to the point that you can barely see the bark of the tree. These are the oaks in cloaks.

Earlier I talked about oak trees as doorways to other worlds. I know practical, down-to-earth foresters who work in temperate rainforests on a daily basis and who still cannot resist stepping through the space framed by a two-stemmed oak tree because it suggests a route to something magical. Alasdair Firth, a man who cares for some rainforest on the west coast of the Scottish Highlands, captures their essence perfectly, if you ask me: 'There is something different about the woods here, even if you don't appreciate it straight away. When you are in them you are kind of inside an organism rather than on the edge of something, because everything around you is alive. The bark's alive as well as the canopy above. There's a lot of different things going on. There's a lot of different colours of mosses and lichens, everything's got its own shade of colour. And it all changes all through the year as well. In winter the woods are still green or grey or white because they have got different things growing on the trees, whereas other woods are mostly brown.'

The UK's rainforests are rich in wildlife. Especially spectacular is that they can host several hundred species of bryophytes and lichens – many more than could be found in any other temperate forest. I'll explain what they are in more detail below. But meanwhile they enhance rainforest's fairy-tale appearance. They upholster the bark of oaks in cushions and wefts, which in turn provide homes to insects, fungi and many microorganisms. Oaks are by no means the only trees offering such plentiful lodging to these species – trees such as ash, birch, hazel and rowan provide rich hunting grounds for such things. Even oak moss – despite its oaky name – is not exclusive to the oak tree. But the oaks stand out for their longevity. The way that oak trees can hang around for hundreds of years means they offer stable and long-standing accommodation. The massive surfaces of big, old oaks also help in the sheer quantity of space they offer. So let's delve further into the worlds of oaky bryophytes and lichens: what are they and why are they so special?

THE WORLD OF BRYOPHYTES

Bryophytes, for the uninitiated, are mosses, liverworts and hornworts – very simple plants. They are 'essentially small, green, photosynthetic plants that do not produce flowers, seeds or fruits', according to Ron Porley and Nick Hodgetts in their book *Mosses and Liverworts*.[16] They may be small, but they are really important. Mosses, for example, are much more wonderful than you might at first imagine. Did you know that moss was what got our planet's plant party going? Moss and its relatives were the first plants on earth, appearing more than 450 million years ago.[17] Kudos to the moss.

Mosses, liverworts and hornworts evolved separately,[18] but share similar talents, for example, holding lots of water. Porley and Hodgetts describe them as 'little land sponges' and as a result these tiny plants can play a key role in calming water flow through landscapes. A high diversity of bryophytes can be an excellent indicator of habitat quality in our Celtic rainforests, reflecting a long ecological history of sheltered, damp microclimate and a lack of air pollution.

Mosses bring great beauty to our world. They are a bit like a fresh snowfall in the way that they make everything they cover look more lovely and more special. Like the oaks themselves, they have links with folklore; in Germanic countries there are tales of Moss People, shy fairies that hide in trees and sometimes borrow things from humans.[19]

And they are woven into stories. The author and ecologist Lisa Schneidau, for example, captures their allure in 'Mossycoat', from her *Botanical Folk Tales of Britain and Ireland*: 'The coat was made of green feathery moss from the woods, and gold thread as fine as gossamer, and nothing else. It shimmered with all the shades of green as living sprigs of moss peeped through the golden threads. The mossy fabric was as light as down, and as comforting as earth, and as flowing as water.'[20]

Mosses are more than just a pretty face though. Some have antiseptic properties, and their spongy qualities have proved helpful in staunching wounds including those suffered during the First World War. In more modern times they have been used for clearing up oil spills. And they can tell stories. When a 5,300-year-old frozen man, nicknamed Ötzi, was found in the Italian Alps, scientists could trace his journey with the help of 75 different species of bryophytes found in and around his corpse.[21]

Bryophytes have great names. And this is where Porley and Hodgetts and I don't quite agree. The authors call some of the names 'just plain silly' giving the example of 'great hairy screw-moss' for *Syntrichia ruralis*. I understand why some people might eschew them but I like the common names; for me they are part of the fun. Why not enjoy the descriptive delights of oak dwellers such as prickly featherwort, frizzled pincushion and shady earwort? Porley and Hodgetts note that Sean R. Edwards wrote a paper on common and English names for the British Bryological Society.[22] There the descriptive delights continue: juicy silk-moss; elegant silk-moss; long-beaked water feather-moss; spotty featherwort.

The UK's cool, moist climate, not necessarily loved by all, makes the country a bit of a European league winner when it comes to bryophytes. Porley and Hodgetts say, 'At the last count there were more than 1,052 different kinds of bryophyte in the British Isles, excluding varieties, comprising 754 mosses, 294 liverworts and 4 hornworts, representing over 60 per cent of the total European flora.' And to put that into context they say, 'We have less than 20 per cent of the European flowering plants.'[23]

One of the great things about bryophytes is that they can be enjoyed in all seasons. We can get to know them – or some of them – fairly easily. And, while they are particularly impressive in the rainforests, I think there is a joy in that some of their species can be found almost everywhere; not only on trees, in woods and on mountains but also in humbler places such as on walls and streets. What's not to love about a species that comes into its own in the rain if you live in a damp place like the UK? With that rain comes mossy magic, softening shapes, muffling harsh sound, painting our world with greenery and poetry.

THE WORLD OF LICHENS

While exploring the species in and around the oak tree, I have fallen especially in love with lichens. They, too, open new worlds. It's not that I didn't like lichens before now. I mean, what's not to like about those little splashes of colour to be found on trees, rocks and other surfaces? But to be honest I'd never really given them that much thought, not beyond the occasional peering down at a pavement, wondering if I was looking at a bit of nature or a piece of squashed chewing gum.

Perhaps you're wondering, what exactly is a lichen? OK. This might take some time, or at least a long paragraph that knowers of the lichen might want to skip. Here goes: lichens have been called 'an ecological strategy'; 'a stable symbiotic association' and 'little forests in miniature'. They have been called 'a marriage of things'; 'a relationship between a fungus and algae' and, as John Craven says in his *Countryfile Handbook*, 'two lifeforms for the price of one'.[24] How does the 'marriage' work? Well, in simple terms, an alga makes food through photosynthesis, like a plant (indeed once upon a time, before taxonomists decided that descriptions of the world needed more than just the Plant and Animal Kingdoms, algae were considered to be the simplest of plants). The fungus, meanwhile, creates a structure on which the alga can live – safe, moist, and with a supply of simple nutrients. In some ways the fungus is like a little agriculturalist, rather than a spouse. Rebecca Yahr, of the Royal Botanic Garden Edinburgh and the British Bryological Society, describes it as 'fungus farming its own population of sugar-producing cells'.[25] Got that? To make it more complicated, sometimes, instead of algae, the photosynthesiser in the relationship is cyanobacteria, otherwise known as blue-green algae. And, just to be really confusing, blue-green algae aren't really algae, apparently. I'm going to leave that one there. In *Lichens*, Oliver Gilbert offers a number of options for defining a lichen, including this fantastically mind-frying one, which, as he says himself, is 'for the pedant': 'A lichen is an ecologically obligate, stable mutualism between an exhabitant fungal partner and an inhabitant population of extracellularly located unicellular or filamentous algal or cyanobacterial cells.'[26] My favourite definition, though, is from David Hawksworth (quoted by Gilbert) who once 'in desperation... at one point suggested that a lichen be defined as "an organism studied by lichenologists"!'[27]

However we define them, lichens are important. Every year, more species are found in the UK, and the last few years has seen our list creeping past 2,000 species. They live in all kinds of places, covering all kinds of surfaces – tree trunks, roofs, rocks, bones, old cars, discarded boots, recycled plastic boardwalk, inter-tidal barnacles – it might be easier to list where they *don't* grow. Many lichens grow less than a millimetre a year but they may live for many, many years, in some cases even centuries.[28] These weird and wonderful little things are big stuff.

One of the most iconic and easily recognised is the tree lungwort (*Lobaria pulmonaria*), also known as 'lungs of the forest', which likes a mature oak tree and looks, well, pretty lungy. Finding it means you might be in an ancient forest.

Other lichen shapes range from crusty tufts to curly kale-shapes to flat splats. They can look like antlers, ear lobes and beards. In each strange silhouette there is a strategy. In his book *Lichens: An Illustrated Guide to the British and Irish Species*, the late, great Frank S. Dobson wrote, 'The shape assumed by a lichen species is such that the algal cells are displayed to maximum advantage for the lichen in a particular ecological niche.'[29]

Lichens have been used in food, medicine and even fashion. Beard (or *Usnea*) lichens, which look like, well, beards – bushy or long and flowing – contain a potent antibacterial and antifungal agent called usnic acid, which has been used medicinally for over a thousand years to treat wounds. I have seen lichen-led recipes as varied as salads, cream puffs and lichen-encrusted beef. Many lichens have been used in dyes. For example, originally the colours that graced the much-sought-after Harris tweed derived from different kinds of lichen on rocks. One of these, the crotal or crottle (*Parmelia saxatilis*), used to give a deep red-brown colour, is sometimes found on oaks, but many lichen species can be used in dyeing, with some species even producing different colours depending on the process used.

In folklore, lichens can be useful in scaring off elves. And they play a key role in well-dressing, the custom of decorating village springs with floral pictures, which is particularly famous in Derbyshire and which appears to be pagan in origin.[30]

Like bryophytes, lichens can tell you a lot about air quality; being without roots, they depend on the atmosphere for their water and nutrition, which makes them particularly vulnerable when the air fills with pollutants. Acid rain (caused when 'atmospheric pollutants such as sulphur dioxide and nitrogen oxides mix with water vapour in the air '), famously a terrible problem in the British Isles in the era of heavy industry, had a big impact on our lichens.[31] Only the most tolerant species such as the grey-green dusty, crusty *Lecanora conizaeoides*, which looks a bit like spray paint, could survive in big cities. The Clean Air Acts have had an impact on these emissions, showing that politicians can make a difference when they are willing, and now a range of sensitive lichen species are spreading back to city centres in particular, and eastern England in general, from their cleaner air refuges in the west of Britain. A politician's work is never done, of course, and lichens and bryophytes can be employed as a kind of canary in the mine. Nitrogen pollutants such as the high concentrations of ammonia given off by some intensive agriculture and nitrogen oxides from burning fossil fuels, for example, are still damaging wildlife. Because lichens, including the beardy *Usnea* types, are so sensitive to this, they can help us read the problem. A report by the charity Plantlife, 'We Need to Talk About Nitrogen', flags that 63% of the UK's most sensitive wildlife habitats are affected by excessive nitrogen deposition.[32] And it isn't only nature that is at risk. So is human health, through reduced air quality, through related contributions to climate change and through the slow destruction of the natural world on which we all depend.

A MOSS THAT IS REALLY A LICHEN

In his book *Lichens*, Oliver Gilbert highlights a good way into the world of lichens: 'Beginners usually start their career in lichenology by investigating woodland or wayside trees... their study may commence by trying to separate *Evernia prunastri* with its matt surface from glossy *Ramalina farinacea*...'[33]

Before I read this, I had by chance taken the very route he suggested, and found it absolutely is a great way to start. In my case it was summer 2021, the sun was hot, high and doing the month of August proud. We were blessedly free of lockdown rules again; once more able to roam the UK and visit its special places. So it was that Professor Brian Eversham and I found ourselves standing in the dappled shade of a mature oak in the New Forest – which is of course an old forest, and is less polluted than many parts of the UK.

Lichens and mosses had draped this tree in rich, varied greens, swathing and swagging right down to the forest floor. This plant could have been dressed by the fairies themselves and the effect made for a most glamorous oak. There was even an attractive smell about it if you knew where to sniff.

Brian knew where to sniff. He raised his hand to a twig and cupped an odd and exquisite entity, a little explosion of grey-green antler-like shapes, all small enough to sit in the palm of your hand.

This was oak moss (*Evernia prunastri*).

There are three things you should know about oak moss:

First, it is not a moss. It is really a lichen.

Second, it has magical properties, or so we are told; magical money-making properties. It is said you can raise funds by casting spells with it. I have even seen it advertised for sale online with advice to rub a bit of oak moss on your wallet before you go gambling to increase your chances of a win. Perhaps we should just pop some in our pockets to help care for our coinage.

Third, the scent. On a delicate rub – we don't want to hurt it – oak moss gives off a faint aroma: woody with a hint of sweetness. The scent might seem vaguely familiar, and that would be because oak moss is a key ingredient in Fougère and Chypre perfumes. This smell is presumably one of the reasons people have valued oak moss at least since the times of ancient Egypt; baskets of it were found in ancient royal tombs.[34]

Our little burst of oak moss was not alone in dressing this pretty tree. Sitting alongside it were many fellow lichens, as well as mosses that were actually mosses. Down at the foot of the tree's trunk, for example, was a particularly fetching sweep of rich, emerald green. It seemed to me a green worthy of Enid Blyton's *Magic Faraway Tree* books, where children and pixies, exiting the magical tree by their spiral slide, plop out to land safely on a tuft of soft moss. Brian knelt down and stroked the mossy base with the gentleness with which I caress Pepe the pooch. At this point I was standing nearby just wondering really what he was up to. He looked up, hand poised on the trunk like a stop signal and said, 'If you see a naturalist stroking a tree it is probably because they are looking for a snail. It lives in the moss and it is easier to find by feeling than by looking.' I had to join in the fun, of course.

The snail in question was a tree snail. It is only around 8 millimetres long, and a beauty with a slender spire of a shell. When you look it up online, up pop interesting words such as 'slightly fusiform' meaning a spindle-like shape that is wide in the middle and tapers at both ends; think 'skinny lemon' for shorthand. Even as a gardener this is a snail I could love. It turns out that stroking a mossy tree is not a bad way to be at one with nature; very meditative. It also probably looked pretty odd to the couple who wandered past. 'It's OK, he's a naturalist,' I called out, hoping this was reassuring.

MINI WILDLIFE WORLDS AND HOW IT ALL LINKS UP

Lichens, like mosses, are great for wider wildlife. They create new habitats, tiny worlds, by providing endless nooks and crannies for ants, mites, moths, bugs, beetles and spiders. They are homes, they are camouflage and they are decorating material.

Several birds use lichens as material to adorn their nests and the master decorator, the neatest of the tweeters, is the long-tailed tit or bumbarrel. Long-tailed tits use nearly 3,000 flakes of lichen, along with moss and cobwebs and feathers on their nests. These adorable, snug homes are the shape of little rugby balls. As an aside we should pay tribute to spiders here. The birds use cobwebs in their building-material mix. The stretchy webs allow a nest to swell to the size needed by a growing family of 6–9 (and up to 14) chicks. If only we could make our own homes out of spider silk, that might solve a lot of housing issues. Our bumbarrels place their lichen patches on the outsides of their nests with the pale surface outwards, taking advantage of the Velcro-like rhizines on the underside of leafy lichens (rhizines look a bit like short roots, but they don't suck up anything useful for the lichen, they are just to attach the lichen to its substrate – or to stick a lichen to a bumbarrel nest). The birds are choosy about their lichens, picking out special soft grey-green lichens from nature's colour chart. These include hammered shield lichen (*Parmelia sulcata*), which likes a bit of oak and is worthy of Farrow & Ball paintwork, although it might need a more snappy name for a colour chart. This décor makes the long-tailed tit's nest one of the most beautiful in the business. Why would birds go to all this effort to cover their nests in this way? There are several theories. One is that the lichen reflects the light, helping the nest disappear into the background. This seems to be supported by the way birds sometimes use bits of polystyrene instead, which sounds much less pretty but, well, I suppose it is a form of recycling.

Some moths also like a lichen. The Brussels lace moth is, says Gilbert, 'closely adapted to a lichen-dominated environment, with a greenish-grey caterpillar that feeds on lichen and resembles a lichen in colour, form and attitude so precisely that it is almost impossible to detect, while the adult moth displays an elegant mimicry of a lichen-covered surface'.[35] In Chapter 3 we saw how the peppered moth can come in different shades, depending on the colour of tree trunks. As sooty pollution levels have fallen and lichens have returned to cities, the proportion of the light-coloured version has increased. Lichens can even have their own galls, tiny ones smaller than peppercorns, made by mites and nematode worms – another teeny part of the jigsaw that makes up the multi-layered world of a tree.

LOVE AMONG THE LICHENS

Back in the New Forest, Brian and I were admiring our glamorous oak tree when something special caught our eyes. A finely angled sunbeam shining down through the leaves had lit up something on the oak's trunk. It was the sign of a lichen-loving organism. We moved in closer and saw the silver path of a special creature, made apparent by the pattern of light through the leaves. In this moment it looked sublime, flash-lit as it was by the heavens.

OK, it was a slug trail. But it was special, I tell you it was special. It told a story of a journey and it was a journey that ended in love. Slug love. Brian explained how we could track the slug, trace its movements. I learned which direction it was going because, part way up the trail, where our love-struck gastropod was obliged to bridge its way from one bit of knobbly bark to the next; the silver trail had a little downwards point. This showed us the trail was heading upwards.

We traced the slug's trip further up the tree to where the trail fanned out messily. And this is where it got amorous. Before we go on, if you could imagine up a bit of romantic music – a spot of Barry White, say – it would help this next bit. 'You've got a lot of slime where two slugs have been going round and round each other prior to mating,' said Brian. You sexy slugs, you.

By way of species, Brian diagnosed a tree slug. How do you know such things? I asked this god of small things. 'Well,' said Brian, 'the only two species that would behave like that, that high up a tree are tiger slug and tree slug. The tiger slug is the one that mates on a rope of slime, dangling down, and although there's a lot of activity there…' Brian was pointing at the slug boudoir, '… that makes me pretty sure they have mated; there isn't a solid mass of slime from which a rope could have extended, so there's not quite enough slime to make it a tiger slug so therefore I'm guessing it is fairly likely it will be a tree slug.'

Now, I know not everyone likes a slug. But they are special in their own way and, I would argue, are quite lovable. And they do their bit for wider nature; as Marion Bryce of Long Eaton Natural History Society says, they are 'nature's great recyclers' and if you can see no other reason to embrace them, they make a juicy snack for a bird, a ground beetle, a hedgehog or a slow worm.

· · · Reasons to love these mini worlds · · ·

Lichens and bryophytes deserve a bit of love, and we need people out looking at them and getting to know them. There are three excellent reasons to get into them.

1. They don't whizz about. Unlike many other species, they tend to stay still when you are trying to observe them.
2. You can find some near you – not only trees but on all kinds of other surfaces.
3. You can enjoy them at all times of the year. They offer an excellent reason to get out into nature in winter.

Plantlife's guides and Field Studies Council guides are helpful, as are online articles from the British Lichen Society and the Natural History Museum. As Plantlife says, 'This is the good bit! Arm yourself with a hand lens and get out...'

A TOP TIP TO IMPRESS

If you're following the beginner's route into the world of lichens suggested by Oliver Gilbert, here's a tip to impress your nature-loving friends: the shaggy strap lichen *Ramalina farinacea* and oak moss *Evernia prunastri* look quite similar on first sight. Both look like little eruptions of shaggy, green-grey strands. An easy way to tell the difference is to lift up the shaggy bits and look underneath – the oak moss is bright white underneath and the other one isn't, as demonstrated brilliantly by Mark Powell online.[36] Perhaps then mention that oak moss is also sometimes called the 'perfumer's lichen' and you are away. Thank you to Brian for this top tip.

. . . . . .

6

Incredible Edibles: Health and Healing

WHENEVER YOU WALK THROUGH A TALL,
DARK FOREST, YOU ARE WALKING DOWN
THE AISLES OF A HUGE GROCERY STORE . . .
IT IS FILLED WITH ALL SORTS OF DELICACIES . . .
PETER WOHLLEBEN, FROM *THE HIDDEN LIFE OF TREES*

WHAT'S EATING THE MARTON OAK?

Ladies and gentleman; birds and bees; beetles, bacteria and bryophytes; organisms one and all. I'd like to present to you one of the greatest plants in the kingdom: the mighty Marton Oak.

The Marton Oak, which sits in a private garden in Cheshire – a sessile – is massive. It is immense – huge – vast – gargantuan. It is also, weirdly, largely not there at all. How so? Well, it is famous for having a gigantic girth and yet most of its trunk is missing. The Marton has been emptied and rotted and hollowed to such an extent that it now stands in four distinct pieces. It is, arguably, now four different trees.

It is being eaten. One of the culprits is not shy about its participation in the feasting. In fact it is looking pretty cheeky on the day I visit. Parts of it are sticking out of the trunk like a set of sassy tongues. Blood-red, semicircular, they look as if they are saying, 'We did this, ha ha!' These are beefsteak fungus (*Fistulina hepatica*), or, to be accurate, the fruiting bodies of this extraordinary fungus.

That name, beefsteak fungus, is well deserved. Its 'tongues' – or brackets – are the meatiest-looking things I have ever seen that aren't actually meat. There is a clue in the scientific name too: 'hepatica' means liver-like.[1] And while it is eating the oak tree – in a manner of speaking – the beefsteak, in turn, is edible... if you like that kind of thing.

We've established that the oak is a much-munched tree. Many creatures like a nibble of it. Meanwhile, some of the species associated with the oak, like the beefsteak fungus, can also be eaten by people and other creatures. For humans, the oak and its acolytes are said to come with a number of health benefits too, albeit some more convincing than others. And when you start delving into the edible species associated with the oak, you discover yet more ways they link up with the wider world both above and – in some cases – below the ground.

Let's take a little culinary tour in and around the oak tree, but first, a health warning. This probably goes without saying but please don't ever even think about eating anything that you aren't 100% sure about. You would be forgiven for thinking the worst case scenario here is death. It isn't. It really isn't. Worse than death is a slow, really nasty death. And worse than a slow, really nasty death is one where you know you are going to die and you know you can't prevent it – when you have been told there is no antidote – and then, in all this knowledge, having a slow, really nasty death. This has happened to some people who have eaten the wrong fungus.

And, while we are on health, let's also think about nature's health and not over-foraging. (See the websites of conservation organisations for more details about responsible foraging.[2]) More brightly, of course, there are many experts working with local conservation organisations to offer forays, courses and fascinating days in the great outdoors.

There is so much going on in the oak eatery that it is hard to know where to start, so let's begin at the beginning: acorns.

TASTY, MASTY ACORNS

Acorns are amazing. These iconic little seeds not only have the power to erupt – slowly – into a beast of a tree like The Marton Oak, they are also – famously – fodder to jays and squirrels as well as mice, badgers and, among other wild things, the astonishing acorn weevil, which in some species has mouthparts almost as long as its own body.[3] Perhaps less well known is that humans can eat acorns too. There is even a word for acorn-eating: balanophagy.

Eating acorns has often been associated with times of famine and desperation, with the poster child being roasted-acorn coffee, the famously derided caffeine substitute of the Second World War. However, William Bryant Logan, in his book *Oak: The Frame of Civilization*, puts forward the idea that the acorn was a staple part of the human diet in some communities back in time. And the ancient Greek poet Hesiod said, 'Honest people do not suffer from famine, since the gods give them abundant subsistence: acorn-bearing oaks, honey, and sheep.'[4]

I recently came across some wonderful 1933 footage of acorn bread making in the United States, courtesy of the Orgone Film Library. In flickery black and white film, against a stunning mountain backdrop, an indigenous American woman referred to as Maggie née Tra-bu-ce (elsewhere Ta-bu-ce) prepares a meal that includes acorns.[5] It offers a few clues as to one of the likely reasons eating acorns went out of fashion. It isn't easy to make your lunch from scratch from acorns.

Ta-bu-ce first gathers her acorns and stores them in a 'chuck-a', which looks like a little hayrick and is made of boughs and pine needles. When she's ready to make a meal, she takes some of her acorns, shells them and then painstakingly pounds them into flour with a rock. She winnows the flour and makes a bowl of gravel and twigs. She filters cold water through the flour to leach out the bitter-tasting tannic acid. She repeats the process with hot water. She does this several times over before she gathers the dough together ready for baking. She even makes her own oven out of hot rocks to bake her hard-earned meal, which she then enjoys with a smile whose warmth has lasted 90 years so far.

I was inspired by this to try a bit of balanophagy myself. I merrily collected acorns during a recent, prodigious mast year, stored them in a nice jar, and looked at them a lot and thought about how I'd one day make some flour out of them. I thought about it some more and then, somehow, a year had gone by and I realised I was probably never going to get around to making my own acorn flour. I bought some online from a lovely-sounding person called Irena, who runs an acorn products website from Slovenia.[6] Irena sent me a family recipe to make acorn bread.

We made it one Easter when I was surrounded by a ready-made gang of food critics – siblings, nieces and nephews. My brother, Mark, came in particularly handy because he actually knew how to knead dough. Nonetheless I fully expected my first attempt to end up in something pretty awful that would give a bit of a giggle to all involved. To my amazement it all went well, really well. The family gang described it as 'earthy, nutty, filling, creamy, mellow' and, slightly more ominously, 'It's got something I can't quite identify in here.' Great-niece Amelie loved it with the rosemary that we had added as a garnish. So, all in all, we recommend it, as a blast of a taste from the past. Eat your heart out *Great British Bake Off* (have we had an acorn-baking challenge on GBBO yet?). Thank you Ta-bu-ce for the inspiration, for opening a door and for that great smile. And thank goodness we didn't have to make our own oven.

You can find the recipe at the end of this chapter, with kind permission of Irena. For more acorn recipes I can also recommend *The Tree Forager* by Adele Nozedar and, of course, a search online. It turns out there are all kinds of fun acorn-based foods to be made, including flapjacks, pancakes, cakes, biscuits, burgers, lasagne and enchiladas. The general advice is that they all tend to be filling, and they are fun as long as you add plenty of other flavours.

Different species of oak provide varying acorn delicacies. I hear some acorns are sweeter than others, so more readily munchable than those of our native species. I have read that the seeds of the American white oak (*Quercus alba*) can even be eaten raw.[7] Logan tells us that the fruits of the evergreen oak (*Quercus ilex* – also called holm oak and holly oak for its pointy leaves) make for a kind of 'festival food in southern Spain and in Morocco, Tunisia, and Algeria'. (The holm oak isn't native to the UK but it grows readily here.) Of the same tree's seeds, he says, 'Up until the early years of the twentieth century, the favourite snack of the ladies at the grand opera in Madrid were roasted salted acorns.'[8]

Meat eaters can get an acorn fix of sorts without all that leaching and grinding business. You can get your livestock to process it for you. I'm talking about pannage, the old practice of letting pigs into a wood so that they feed on acorns and other tree seeds.

William Cobbett wrote on pannage back around the nineteenth century: 'The only good purpose that these forests answer is that of furnishing a place of being to labourers' families on their skirts; and here their cottages are very neat, and the people look hearty and well, just as they do round the forests in Hampshire. Every cottage has a pig or two. These graze in the forest, and, in the fall, eat acorns and beech-nuts and the seed of the ash; for these last, as well as the others, are very full of oil, and a pig that is put to his shifts will pick the seed very nicely out from the husks.'[9]

Leaving aside the comment that this is 'the only good purpose that [some] forests answer', if you eat meat it would seem to be a good plan to put a pig 'to his shifts' to turn your acorns into bacon.

Poison alert: While I don't imagine you'll be planning to fatten up other types of animals in this way, please be cautious about too much mixing of animals and acorns. I understand they can harm some creatures if too many are ingested.

Other parts of the oak, beyond the acorn, have been brought to people's tables throughout history. Do some very careful homework before you start experimenting, though, and exercise caution. Acorns are rich in tannin, for example, and, again, too much of that can be bad for you.[10]

NIBBLING FURTHER INTO THE OAK

Some people join the caterpillars in chewing on a young oak leaf or three in the spring. I have had a go myself with a few very young leaves, which proved to be an interesting, slightly bitter addition to a salad. Oak leaves do need to be very young to be palatable; they quickly become tough and very bitter – all the better for putting off the oak nibblers we met earlier in the book.

More familiar to most is the practice of using oak wood to add a bit of extra flavour to a meal. You might enhance your dinner by cooking it over oak charcoal or smoked oak shavings. In his book *Plants with a Purpose*, Richard Mabey has some advice on using different kinds of wood for smoking: 'Normally it is just a single species of hardwood, selected because of the affinity between its scented smoke and the food: oak for ham, birch for haddock, apple for oysters.'[11] He quotes John Wyatt in *The Shining Levels*, to tell us: 'Old oak has an honest, pungent lusty smell as you would expect.'[12]

Perhaps for dessert sir or madam might like to rekindle the culinary adventure again by enjoying a chew on a sweet gall; a nice young oak apple perhaps, as was enjoyed by some people in the past – not one for the veggies this, what with the wasp larvae and who knows what other tiny insects in there.

You could perhaps accompany your oak-themed meal with a nice little glass of wine, or how about a tot of whisky if you are in Scotland – whiskey if in Ireland – all oak-aged of course.

A BARREL OF FUN

Back in 1915, David Lloyd George ruled that whisky must be matured in a cask for at least two years (this was later increased to three years). Perhaps surprisingly, Lloyd George was a teetotaller at the time. It seems he was looking to water down the drinks industry, yet his ruling inadvertently secured the reputation of whisky as a premium product; before then, some producers were selling iffy, immature, unwhisky-like versions of the drink, damaging its reputation.[13]

So what is a cask? Butts, barrels, firkins, pins, hogsheads: these are all casks of different sizes and oak is a wood of choice in which to age whisky and wine (even some cheeses get to spend a bit of time hanging out in an oaken box). I should point out right here that little of the oak used for casks in the UK is home-grown these days; many of our casks come from the US and Spain, but barrels were big business in the UK back in the day, important for liquids and for keeping provisions away from unwelcome nibblers. They are also things of great craftsmanship; they need to be perfect of course – I can only marvel at the skill involved to make something from several pieces of wood that doesn't leak. And then the further skill of choosing and treating the oak so that its compounds emerge as a hint of other joys – vanilla, butter, spice…

While casks date back to ancient Egypt, 'it is generally regarded that the barrel with staves originated with the Celtic peoples in northern Europe', says whisky writer and consultant Charles MacLean on BBC's *The Food Programme*.[14] I'm just going to pause for a moment there to let that job title sink in – whisky writer and consultant – in case there is anyone out there considering their career options right now … Back to Charles: 'By having a belly on the vessel you can roll it and manipulate it much more easily than if it was a box, for example. They can be spun around; they can be moved around; they can be racked.' He says the best barrels were made of oak. Why oak? It is strong and malleable – you can bend it into shape (after heating or steaming) and it is fine grained enough to keep the liquid in while also allowing it to breathe a little. From the latter comes the lovely concept of 'the angels' share', the drink that evaporates through the walls of casks and which has lent its name to several bars, a hotel and a film.

Thankfully, given all the work involved, casks are recyclable. A whisky/whiskey barrel might have enjoyed a previous existence as a bourbon barrel.[15] And while a posh wine might be given a brand new oak barrel, others might be aged in a mix of old and new. Ultimately, casks will be turned out to pasture – perhaps in a garden like ours: we have an old barrel that serves as a butt to collect rainwater. Even after a decade in our garden, every time you open the lid it gives off a faint aroma of whisky. I like to think of that as our own little 'angels' share'.

OAK THERAPY

From supermarket to pharmacy: over the centuries the oak and the species it supports have acted not only as a larder but also as a source of cures for various human conditions.

Certain oak trees have been particularly celebrated for their healing powers. People are said to have taken bark from The Marton Oak to rub on warts, rashes and boils, or hung it in their homes to ward off the 'Evil Eye'. Meanwhile the ancient Remedy Oak, near Wimborne in Dorset, is believed to have provided a venue where King Edward VI engaged in a spot of healing. Being a king at the time apparently gave you special powers to cure the sick, including those who suffered from 'King's Evil', or scrofula – a swelling caused by tuberculosis. Edward's therapy session apparently took place in 1552. Unfortunately, he died a year later – probably from tuberculosis.[16]

Marble galls have been pressed into service to cure piles – just mix with hog's lard and apply to the offending area (I probably wouldn't try this at home), while the oak's leaves, bark, branches and acorns have been sought out at various times to cure diverse complaints. These include warts, rashes, pimples, boils, diarrhoea, ringworm, sunburn, eczema, kidney stones, sweatiness and freckles. Freckles! What is wrong with those little facial sunshine spots? I've no idea, but anyway spring dew from oak leaves is said to do wonders for the complexion.

If you were to list the remedies attributed to the oak by lack of credibility there would be some strong contenders for the top slot. The winner might well be the toothache treatment based on hammering a nail into an oak trunk – you have to pity the poor trees that have taken a hit for the team in the name of that therapy. And it isn't easy to see how carrying acorns promotes fertility and slows down the ageing process. On the latter I guess you could make a case for the nutritional benefits of the acorn-eating boosting one's health, or indeed for having an acorn on your person for the sheer childlike joy of it, which in itself may help keep you young at heart, but I'm not sure an oak seed, powerful as it is, is going to bring us the full Benjamin Button experience.

One of my favourite claims for the health-giving powers of the oak was recorded by the diarist John Evelyn, who noted that sleeping beneath an oak tree would not only cure paralysis, it would also 'recover those whom the malign influence of the Walnut-tree has smitten.'[17] That might be a bit harsh on the walnut.

Could any of the health claims for the oak's powers be true? Well yes, almost certainly on some levels. Take the custom of using oak twigs for toothbrushes, for example: the idea of using a bit of tannin-rich bark doesn't seem too off the wall. Fiann Ó'Nualláin, in his *Holistic Gardener* series of books, says tannins are wound cleaners and that 'the term tannin comes from the High German tanna, denoting oak or fir trees so you can make a rinse from some oak or pine bark'.[18]

Parts of the oak, including its bark, have been attributed with anticarcinogenic qualities, although, as touched on above, I have also seen claims that high intakes of tannin could be linked to some cancers.[19] So, as with all these claims, it is vital to tread with caution.

Meanwhile the chicken of the woods fungus – which often grows on oaks, and plays a similar 'rotting' role to the beefsteak fungus, hollowing out the heartwood – appears to have all sorts of interesting medicinal qualities.[20] I have read claims that it has antimicrobial and antioxidant properties, among other qualities.[21] I have also heard of it being used as an insect repellent to ward off flies, midges and mosquitoes.[22] What these qualities suggest to me is that the oak and its world have a multitude of surprising and interesting properties, some of which we understand, and others we partially understand or – possibly – have yet to discover.

More broadly, everyone who has been in a woodland or sat beneath a tree knows they feel a bit better for the experience so, on that basis alone, Edward VI's patients might well have left his 'surgery' feeling a little healthier. We know that time in nature in general is good for us both mentally and physically. Trees, of course, clean the air and give off oxygen. Some studies suggest that walks in forests can 'increase pulmonary function' and 'reduce arterial stiffness' and that these benefits can last for several days. People are talking increasingly about the healing properties of chemicals given off by plants. Claims are made for their effects on our well-being, even to the extent that they could have healing effects on serious diseases such as cancer. In Japan the idea of forest bathing – *shinrin-yoku* – is well established and involves a natural form of aromatherapy that comes from being in the forest and taking in its air.[23]

I'm always glad to hear of schemes that help people harness nature to help themselves. One such is Resilient Young Minds, a collaboration involving the National Trust, the Woodland Trust and the National Lottery Heritage Fund. Young people gather in Fingle Woods, Dartmoor, famous for its ancient oaks, the type with boughs 'mossed with age'. In an online video introducing the pilot, experts including Dr Emma Chapman, GP and clinical director of the North Dartmoor Primary Care Network, says, 'The aim of the project is to connect people not only with nature but to build on skills such as resilience, mindfulness and teamwork, skills that they can translate and take forward into their life.'[24] In the film the young people chat happily and openly about their feelings and mental health. When they first arrived they were described as 'nervous' and 'anxious'. They now talk about enjoying the birdsong, and the sound of the river, saying things like, 'I feel safe', 'I feel myself', 'I have gained confidence'. The change isn't only because of the time they've spent in the woods, of course, it is also about the sharing of skills and support, but the key thing for me is that using the woodland as a venue for social prescribing such as this makes so much sense.

I am heartened to see public figures coming forward to talk about the way trees and nature can improve mental and physical health, whether you are having a hard time or could just do with a bit of a boost. The broadcaster Dan Snow, in Adam Shaw's *Woodland Walks* podcast series, says he reached a moment in his life when 'I started to realise at that point that the woods were a place of extraordinary healing and balm and importance to me and my wellness. And so no, I'm not walking around these woods and thinking about the oak trees around us that would have been important in the construction of Nelson's navy in the late eighteenth and early nineteenth century, although occasionally I see a branch and think that would make a good part of a ship's hull... I'm mostly just being really happy that I'm in the woods and I'm regenerating.'[25]

Doctor and TV presenter Amir Khan says: 'The effect of nature on our mental health is really powerful because there are real physical changes going on inside your body ... When you're looking at birds, at any kind of green space, you produce less stress hormones and your blood pressure and heart rate drops.'[26]

If anyone was ever in doubt of the gentle healing qualities of nature and trees, oak and others, I would recommend simply getting out there and having a go at it. A simple test is to rate how you feel before going out on a walk on a scale of 1 to 10. Then do the same after a little walk, ideally while sitting under an oak tree. I can't 100% guarantee you'll feel better, but I *almost* can.

THE WIDER WORLD OF THE OAK EATERY

There is no shortage of culinary celebrities among the species supported by oak trees. Many of the big stars are fungi, or rather the fruiting bodies of them – the mushroomy-type parts. Here are a few just as a taster. And please, please don't forget that health warning not to try anything you aren't 100% certain about.

Boletes for eats

Many of the bolete family love an oak tree, and we humans love many a bolete to eat. The summer bolete (*Boletus reticulatus*, sometimes called *Boletus aestivalis*) is a welcome sight early in the mushrooming season, and we'll explore its amazing underground world shortly. It is one of the boletes that are almost of a kind with the famous penny buns (*Boletus edulis*), which we also know as ceps or porcini.

Grisettes and a deadly warning

The grisettes might sound like a 1960s girl band but they are arguably a bit more B-list on your plate. These are the Amanita fungi (*Amanita* sect. Vaginatae). Not for your beginner forager, these, as they can easily be confused with other, toxic Amanitas, including the deeply deadly deathcap (*Amanita phalloides*), which can also hang out around oak trees. You really don't want a mix-up with a deathcap. It can take half a cap or less to kill a person, which is why they have been the murderer's mushroom of choice for several centuries. The Roman Emperor Claudius was – allegedly – killed by his wife, Agrippina, after she mixed deathcap juice with Caesar's mushrooms (*Amanita caesarea*).

Chicken of the woods
The chicken of the woods (*Laetiporus sulphureus*) earned its way onto this list because, one July, it suddenly, joyfully, popped into view out of The Little Owl Oak in our village. It looked like layers of foam filler had spilled enthusiastically out of the insides of the tree: a lovely, happy, bright yellow sight – it doesn't have 'sulphureus' in its name for nothing. It is sought out by food foragers for its poultry-ish taste. Note some people – I have seen estimates for 20% of the population – are allergic to chicken of the woods. Not to be confused with the also edible, also named-in-a-poultry-inspired-way, hen of the woods (*Grifola frondosa*).

The beefsteak
We met the meaty beefsteak fungus earlier when we met The Marton Oak. Some people say the young brackets of this fungus – those cheeky tongues – taste like a cross between beef and melon. However, the foraging expert John Wright almost left it out of his book *The Forager's Calendar*, saying it tastes like rubbery, under-ripe tomatoes. Never mind, I'd argue that the beefsteak is best left on the tree anyway, to do its cheeky thing sticking out its tongue at passers-by.

The summer truffle

The summer truffle (*Tuber aestivum*) makes the list really by being an A-lister in the fungal world, beloved as it is by gourmets. To find it you might need to go underground. You might even need to enlist the help of an animal friend, such as a specially trained dog or pig, to sniff one out. I understand that most dogs can be trained to detect them, so perhaps there is a chance our Pepe the pooch could be gainfully employed for once to earn his keep. Ah the life of the great snoutdoors.

DINNERS AND DINERS

The above list is just a small sample of the 108 fungi highlighted by PuRpOsE[27] that have an extraordinary, intimate relationship with our oak trees. When I say 108, I should point out they are they are just the fungi logged and documented as part of the exercise. There are almost certainly a great many more.

We know that fungi live all over the tree, inside and out. They are in its bark, in its buds, in its leaves, in its leaf litter and on its roots. Fungi help oaks to live; they help them to die; they help them to drink and take in nutrients. Some of them are stars of our dinner plates, others are deadly poisonous. And they were dedicated recyclers way before the word was even dreamt up by we humans. Many fungi have a big impact on the oak's life cycle. Those such as our friends the beefsteak and the chicken of the woods eat out the centre of old oak trees such as The Marton Oak and The Bowthorpe Oak, helping to create new worlds and new homes for all the species that love a hollow tree and all the varied conditions it offers.[28]

You might think that eating a tree's heart is a bit brutal, but it is all part of the oak's clever recycling system. The

Woodland Trust's Jim Smith-Wright says, 'As the tree ages, the wood at its centre is no longer useful to the tree, so that is a massive store of nutrients held inside the tree that the tree can no longer access. So, what some species of fungi can do, is they can start to break this down again, which the tree can't do itself. It can convert it into compost, essentially, and the tree can then live off that compost, so live off itself (in some cases the tree can produce new roots inside itself to live off that compost).'

Professor Lynne Boddy, professor of microbial ecology at Cardiff University and author of *Fungi and Trees: Their Complex Relationships*, explained why gouging out the oak's innards isn't the problem you might imagine it could be for our tree. 'If you think about the growth of a tree, it is putting its extra growth on the outside. There are very few living cells, if any, in heart wood, generally it is already dead. In a big oak tree it is only the outer few inches that are conducting water.' So the useful bits are now on the outside and the innards are now, well, something the oak can recycle.

It's not easy being a rotter though. Despite the incredible numbers of fungi in the world (I have read estimates of more than six million species – and other estimates that go higher), very few take on the task of rotting out the oak. Our tree's heart wood is acidic, and contains tannins and other compounds that make for trying working conditions, deterring other fungi. So we might congratulate our beefsteak on its skills. As an aside I should also mention the beefsteak's achievements in the world of oaky art, as it can give the wood a rich colour known as brown oak, much appreciated by furniture makers; if the tree is only partially colonised, the result is 'tiger stripe oak', a kind of wood that highlights ways blemishes can be beautiful.[29]

The advantages of being hollow don't end with recycling. A hollowed-out ancient tree also has some structural advantages over a solid tree. Becoming a cylinder can help you bend rather than break.

Some trees, such as The Marton Oak and The Parliament Oak in Nottinghamshire, have been hollowed to the point where they look less like one ancient tree and more like several younger trees. In a way they are just that; they are, in a sense, young trees – regenerating by themselves. They are able to do this because, as ecologist and ancient tree expert Jim Mulholland told me, 'Wood is like a collection of straws. Each tube has its own roots and leaves. Over time these can separate from their neighbours. By this mechanism as trees age, they begin to function less as one tree and more like a collection of trees.' So a tree can reorganise itself. I asked Jim if this meant that, in this way, with incredibly favourable conditions, a tree could live forever. He agreed.

This idea of tree immortality made me happy. It also begged the question: at what point does a single tree technically become more than one tree? And I loved Jim's response: 'It depends how you try to define what an individual is. Human concepts of what is an individual fall short when it comes to plants because they can clone themselves. Are two genetically identical trees one or two trees?'

Sometimes it can be fun and happily mind-bending to think less like humans and look at life, as much as we can, through the lenses of other species.

GOING UNDERGROUND
The more I learn about the world of fungi the more amazing I find it. Take our summer bolete, for example. It not only feeds us but, in turn, also plays a key role in sustaining

the oak by serving it water and nutrients. It does this in a slightly odd, controlling way.

This process is hard to see. It involves going deep: down and under the ground. For this we will need guides to the underworld. First up is Dr Andy Taylor, who goes by the imposing title of Molecular Fungal Ecologist of the James Hutton Institute. He's also a fun guy; the kind of person you just want to parachute into a classroom of kids or – even better – into the middle a forest school, because knowledge pops out of him like spores from a ripe puffball mushroom.

Andy tells me the summer bolete is a mycorrhizal fungus. The *myco-* prefix indicates that the term relates to fungi, and *rhiza* means root. It is a logical if not particularly memorable name, albeit one we are increasingly using, thanks in large part to authors such as Suzanne Simard and Peter Wohlleben, who have helped open up the world of mycorrhizal fungi. It is logical because it does what it says on the tin. Key parts of these fungi live on the ends of the oak's root tips, covering them. But there's more: in this case, the fungi in question take us a step further in terms of linguistic gymnastics. They are actually ectomycorrhizal, *ecto* meaning 'outside', as they are *between* the root cells rather than *inside* them. 'They form what's called the Hartig net, which holds each of the outer root cells in a basket of hyphae – and this is where exchange of nutrients and sugars takes place,' says Andy.

The oak employs these fungi to resolve a curious tree challenge: 'The oak tree has an amazing, big root system, but the vast proportion of it is tied up in anchoring the tree into the ground', says Andy, 'now, this same system also transports nutrients and water and, if you are a huge tree, the last thing you want to do is to send something twenty to thirty metres only to lose it through seepage or evaporation, so roots are designed to stop the loss of water.' This, coupled with the need of the root system to prevent diseases getting in, means most of the roots' surface isn't permeable: things can't pass in or out, not water to drink, nor many of the minerals the oak needs from the earth. 'The tree is isolated from the soil,' says Andy, and there's the rub. The mighty oak, king of the forest, micro-ecosystem, can barely help itself even to a simple drink, never mind other important parts of its diet. It needs help. So here's where the between-the-root-cells fungi like the yummy summer bolete come in.

'The parts of the root system involved in the uptake of nutrients are the small, thin roots in the last two or three millimetres,' says Andy. 'They are called feeder roots and they are almost all (95–100%) completely covered by the fungi. If you can imagine each one is wearing a sock of fungi, it is like that: it is called a mantle or fungal sheath. Nearly all water and nutrients go into the tree through this mantle.' This means that oaks are pretty much at the mercy of fungi to supply these things. Andy says, 'The fungi control what's going in and out of the tree. Moreover, each time the root produces a little side branch, the fungi are onto it. They cover the root end, taking control.' These fungi can protect the tree from heavy metals and pathogens.

'It is a bonkers system but it has huge benefits,' says Andy. I nod and start thinking about an oak with its own set of personal waiters living at its root tips, looking after it, bringing it water and mineral nibbles: 'Another wafer-thin slice of phosphates for sir?' Meanwhile the oak pays its fungal feeder back, tipping it not in cash but in sugars such as glucose. These sugars are the food the tree makes from carbon dioxide, sunlight, and water; the photosynthesis done by all those hard-working leaves (those that haven't been eaten by caterpillars, and other munchers we mentioned earlier). It is an expensive process: as much as 30% of the sugars a tree makes goes to feed the fungi. That is a lot of the tree's energy.

It is symbiosis, and 'it is just like you', says Andy, 'you are just one big symbiosis!' I've been called many things in my life, but never a big symbiosis. I try to process this. What is a symbiosis like? Who and what is in my symbiotic community? Apparently there seem to be more bacterial cells in our bodies than human cells.[30] 'You are the minor component of yourself,' Andy tells me, grinning at my expression as I realise I had no idea who I am, who we are, what we are.

This symbiosis thing doesn't end there. When you start working through these chains of thought, you eventually come back to the way we humans rely on plants, including oaks. We breathe their lovely waste product – the oxygen given off during photosynthesis – and they nourish the soils that feed us, for example with their leaves. Most plants need fungi to exist, and from there it is logical to see the extent of our reliance on the fungi that work with the plants and all the other life forms. 'We are here because this partnership between plants and fungi exists,' says Andy. 'It isn't about plants getting established then the fungi come along and help out. This is a vital partnership. Without fungal help most plants would not survive. We would not eat. And, ultimately we would not breathe; basically without this we wouldn't be here.'

Fighting talk
Just when I was beginning to think this symbiosis sounded quite pleasant – one organism helping another in a nice,

supportive network – Andy disabuses me of my friendly fungal idyll. The conversation takes a darker turn. It's not all sweetness and light under the ground after all. 'We tend to think about this mycorrhizal system as being a mutualism, but a mutualism might suggest an equal exchange,' says Andy. 'This is more about mutual exploitation. Symbiosis is about two or more organisms living together; it isn't necessarily about benefitting each other. The organisms involved can be friendly, but they can also be pathogens bringing disease or destruction. They are often locked together in a delicate balance in the game of life.'

Andy is seriously upping the fungal-oak jeopardy here. So who – which, what – I wonder, holds all the cards in this game? Is it the fungi, because they are controlling the water and nutrients? Well, not necessarily. Andy explains that the tree has some top trumps to play too: 'The oak, in turn, has control over how much sugar it gives to root tips, and so to the fungi. It is all a fine balance and it is also a battle. From laboratory experiments we know that, if the plant can avoid giving 15–30% of its sugars to the fungi, it will do so. If we give the plant all the nutrients it needs through fertiliser then it doesn't need fungi, and doesn't develop any mycorrhizal associations with fungi. On the other hand, if you keep giving a mycorrhizal fungus lots of sugar – lots of carbon – it may start to harm the plant. What develops in a healthy natural environment is that each entity gives the other only as much as it needs to keep the relationship going. Knock out the balance and things can get dirty.'

This isn't the only fight for survival. There are other contests going on between the very fungi themselves. It's beginning to sound like *Game of Thrones* down there. 'There might be a couple of hundred fungi sitting on the root system, and the adjacent tips can be colonised by different fungi,' says Andy. I think back to his image of a sock of fungi, except now in my mind this isn't necessarily a neat piece of footwear made of one kind of thread; it's a more complex sock composed of different elements, your intricate Fair Isle pattern perhaps, as opposed to your basic monocolour knit. I ask Andy why the oak wouldn't just stick with the one fungus, keep things simple, relationship-wise, working only with the easy-to-love summer bolete. 'This is probably about insurance against different conditions, changes in temperature for example,' says Andy. 'You have this complex, diverse community with different strategies all taking up water and nutrients into the tree. They have different strategies for foraging but all involve nutrient uptake by the soil.'

Forester Peter Wohlleben talked about this in *The Hidden Life of Trees*, his book that helped bring the world's attention to ways trees communicate: 'In oak forests alone, more than a hundred different species of fungi may be present in different parts of the roots of the same tree. From the oaks' point of view, this is a very practical arrangement. If one fungus drops out because environmental conditions change, the next suitor is already at the door.'[31]

I ask Andy if the same systems are going on in all trees. He explains that oaks aren't alone in having this ectomycorrhizal system; beech and birch have a similar arrangement. Other trees such as ash have a slightly different system, still mycorrhizal but less 'ecto'. They have fungi at their roots, they still engage in this underground networking but, in the case of the ash, the fungi work from within the root cells.

It isn't only about the oak tree of course. There are probably millions of species of fungi in the world, and they are involved in countless interactions. Fungi can be used to make food, medicine and even building materials. All are important, and there is so much to be discovered in their world. Some species were last seen more than 50 years ago. Fungi are under threat from many quarters including habitat loss and pollution. And there is still a lack of knowledge and understanding surrounding their lifestyles and requirements, Lynne told me. 'We know so very little; one of the problems is so little money is put into the study of these for research yet they are the lifeblood of our planet.'[32]

I wonder if they could also help save our planet; help with climate change for example. Lynne tells me there is research to suggest that the 'choice' of fungal partners is changing – we know this from the sightings of the fungi around trees. This leaves me wondering if this change of partners could potentially take place more quickly than the evolution of the oak tree, and could that, in turn, help the oak to cope with the vicissitudes of the world? On the face of it, it would make sense to have partners in life that can adapt more quickly than the trees themselves, working to deal with external threats. As we know full well, trees' lives are long – oaks' lives can be very long indeed – and so evolving takes time. Meanwhile the world is moving on quickly, what with climate change and pests and diseases still coming to the UK. When you look at it like this, having a bunch of beings helping you adjust to threats makes a lot of sense, competitive as that relationship may be.

Hubble, bubble, soil and trouble
Back in Sherwood Forest after one of our many visits to The Medusa Tree, my husband Toby and I sit on a bench, Pepe at our side, and think about the things going on beneath our feet; this whole world we hardly, if ever, get to see, or rarely think about. We now know that hyphae, the long, branching, threadlike parts of fungi are snaking about beneath our

feet, seeking out food. 'In a gramme of soil – less than a teaspoon – you can get 400–600 metres of them,' Andy told me. That is an awful lot of fungi. While an oak's roots have been found 30 or more metres from the trunk – way beyond the leafy canopy overhead – mycorrhizal fungi will extend that reach even further into the soil around them, spreading out and around the roots towards important nutrients and water, slowly creeping through the soil beneath our bench. In terms of reach they are immense, but we still don't know the full extent of that reach.

Botanist Ray Woods compares the underground networks of fungi to the networks of pipes and cables that support a busy city, in *Wood Wise* magazine. 'As with manmade electricity and water infrastructure, fungi are the underground conduits that link and support woodland life.'[33] His memorable imagery highlights why we should be kind to soil. Imagine if someone drove a plough through the plumbing system or the electricity cables outside your home or trampled them to the point of damage. You'd be hopping mad about the damage to your support systems, yet this is what we often do to trees.

Our little underground friends are looking after us in their own way. This in part explains why ancient woodland, particularly, is so important and why, now it is so rare – covering only 2.5% of the UK – we should never destroy or damage it.[34] It is not just about the trees. It is also about the fungi, and the wider system within the soil, all the things underground living in an intricate, delicate, balance.

These ancient woodlands and their soils and wildlife have co-evolved for thousands of years, creating diverse, distinctive and valuable ecosystems that cannot be recreated. And other soil? If it is somewhere else, not in our precious ancient woodland, does that matter? Well yes, it does to your street tree, your hedgerow tree. Imagine the damage their roots, their fungal systems face on a day-to-day basis. Soil needs pores, pockets of air, to function healthily yet so often the ground above the roots of a tree are trampled by people and vehicles. Think of the poor hedgerow next to a field that gets ploughed to the very edges.

On the other hand, if you have a balanced ecosystem, you also have stable carbon that stays in the ground for thousands of years. Look after our trees and forests and we have a natural carbon storage system, which we all know is vital as we work frenetically to avoid climate disaster.

Hyphae as hunters ... and hunters of hyphae
Back at our bench in Sherwood, I peer at the soil, trying to imagine what is going on down there.

I have often wondered why some people say fungi are more like animals than plants so, later, I put this question to Andy. He explains that, unlike plants that make their own food, fungi are a bit like us in that they have to find food from outside sources. I imagine the hyphae snaking around slowly beneath our feet and Pepe's paws. Fungi do, of course, have their own kingdom and they are getting on with things in their own way. So, does anything eat the hyphae themselves? 'Yes they are grazed continually by soil mites and springtails,' says Andy. Springtails. Well there's another whole world in itself. You can have tens of thousands of these tiny, springy, pingy insects in a square metre of soil and leaf litter.[35]

In an article in the *Guardian*, George Monbiot wrote: 'After two hours examining a kilogram of soil, I realised I had seen more of the major branches of the animal kingdom than I would on a week's safari in the Serengeti. But even more arresting than soil's diversity and abundance is the question of what it actually is. Most people see it as a dull mass of ground-up rock and dead plants. But it turns out to be a biological structure, built by living creatures to secure their survival, like a wasps' nest or a beaver dam.'[36]

Back on the bench on another occasion I try to imagine 40,000 springtails and all the other species down there. There is so much going on that we don't see. Pepe, meanwhile, is more interested in a tea bag someone has left behind. It is a small piece of litter, presumably left by a nature lover having a nice cup of tea on the bench, not thinking about the intricacies of life below, of the chemical and bacterial interactions and how a new addition might change or interrupt them. We pick up the teabag and take it home to compost the innards. It is one small thing we can do for the trees, but when you think about it, it is the small things that count; the countless small things that count.

... Bread rolls with acorn ... flour and chia seeds

Here's Irena's recipe, reproduced with her kind permission. I have added my comments in square brackets. We have also converted the decilitres (dl) in the original recipe to millilitres (ml).

- 330 g of white flour [we used plain, white]
- 70 g acorn flour
- 20 g yeast
- 1 teaspoon of sugar
- salt to taste
- 2 tablespoons of chia seeds
- 200 ml warm milk [we warmed it to 'hand hot']
- a little cold milk for greasing [we used about a teaspoon]
- 100 ml water
- a tablespoon of olive oil

In a large container sift flour. Separately sift the acorn flour and add four tablespoons of warm milk, mix well [we used a wooden spoon] and crush larger lumps if necessary. Add moistened acorn flour to the white flour, mix well. In the middle we make a hole into which we pour a good dl [we translated this as 'dollop' in this case and added about a third of a cup] of warm milk into which we put the yeast; add a spoonful of sugar and sprinkle/cover it with flour.

Let the yeast prepare [having dry yeast, we mixed it with the sugar and some of the water 45 minutes before using]. In the meantime put in a cup two full tablespoons of chia seeds and pour over 1 dl [the 100ml] of water. When the yeast is ready, quickly mix it with the flour and add all the other ingredients. So, add the rest of the milk, salt, a tablespoon of (olive) oil and moistened chia seeds. Now knead the dough well for a good (at least) 5 minutes. Then let the dough rise (about 40 minutes) [we did this in a warm airing cupboard]. Re-knead.

We form a loaf that is cut into approximately equal parts [we made two]. These parts are then shaped into bread rolls (buns), which are then halved by pressing down with the handle of a wooden spoon. Put them on a baking tray lined with baking paper and brush [the greasing] with cold milk. Let them rise for about 30 minutes.

Put the prepared buns in a hot oven and bake at 220°C [gas mark 7] for 20 minutes.

Cool and serve. Due to the lower content of acorn flour, which is about 18%, the buns will be very tasty for your children as well.

Have a nice day. Irena.

· · · · · ·

7

Threats: Could a Beauty Be a Beast?

MAY NEVER SAW DISMEMBER THEE,
NOR WIELDED AXE DISJOINT,
THAT ART THE FAIREST-SPOKEN TREE
FROM HERE TO LIZARD-POINT.

ALFRED, LORD TENNYSON, FROM 'THE TALKING OAK'

DEATH AND DESTRUCTION

There is a beetle that shimmers like an emerald with hints of blue and gold. On its slender, tapering tail are spots that gleam like white lights. This creature is as beautiful as a brooch. Is it also a bit of a villain?

The oak jewel beetle (*Agrilus biguttatus*) has been blamed for playing a role in acute oak decline (AOD), which is weakening mature oaks across the UK, particularly in the south. This is causing concern about our mighty trees. But could this beastie be wrongly accused? We will explore our beetle shortly, but first we should note that, sadly, acute oak decline is not the only threat to our beloved trees.

Action Oak, an initiative to protect our oak trees, says, 'Our iconic oak trees face a fight for survival against pests and diseases that have the potential to devastate the oak population.'[1] Oak declines – where we see a drop in the vitality of oaks in a given area – are worrying. AOD is one such, or rather, 'a subset of oak decline, with a specific set of symptoms which are collectively considered to cause rapid decline and death over three to five years,' says Rebecca Gosling, conservation evidence officer and tree health expert at the Woodland Trust. These oak declines sit alongside other threats including climate change; mammal browsing; pollution; oak processionary moth (OPM); powdery mildews and insensitive development, when we chop up our landscape and break the links that connect the chains of life.

More dangers lurk further afield, not yet on our shores, but nearby. Too close for comfort are diseases such as that caused by *Xylella fastidiosa*, which damages plants including oaks by blocking their vessels and restricting the flow of water and nutrients.[2]

The hazards pile up and amplify each other. And they don't only menace the oak itself. As we have seen, any risk to the oak has a knock-on effect on a raft of other species: beetles, butterflies, bryophytes and beyond. And, as Action Oak says, 'Losing oak trees from our landscape will impact our well-being, economy, environment and the species that depend on them.' Ultimately, threats to the oak are threats to our wider natural world, and we don't have to travel too far along the line of reasoning to see that this also means threats to us humans.

We know the oak is a great survivor. In previous chapters we have seen how our oak trees deal with assaults from assailants such as caterpillars and fungi, and can often work out a kind of natural balance with many of them. In his book *Woodlands*, the historical ecologist Oliver Rackham writes, 'Trees, being long lived, are parasitised by a huge variety of viruses, bacteria, fungi, mistletoes, nematode worms, mites, insects and other trees... Trees are damaged by frost, drought, waterlogging, pollution and browsing or gnawing mammals, or altered by genetic mutations. Almost any big tree can be called diseased.' Rackham goes on to say, however, that 'most tree diseases are apparently trivial'.[3] Our trees have, after all, evolved to live alongside other native species – even those that can be damaging at times – and find a balance. The problem now is that the threats are so many and varied – and some of them so new – that our oaks are struggling to cope in places. Evolution, though amazing, takes time.

So here we are, the sad chapter in the oak story when we dive into the bad stuff. We'll roll around in a few of the issues a little – highlight some of those that are key – possibly get a bit depressed, perhaps even throw up our hands in despair for a moment. But before we wallow too much, I promise I'm not going to leave us there, not in this book. If I didn't have hope I wouldn't be writing this. There are so many things we can do to help our wonderful oaks, lots of things; actions big and actions small. A great many of these are not only important to do but they also are fun (I have listed some at the end of this book). But for now let's have that wallow. To start with, let's take a closer look at our beetle beauty – could it really be a bad beast?

THE OAK JEWEL BEETLE

The oak jewel beetle is also called the oak splendour beetle, the two-spot woodborer and the two-spotted oak buprestid.[4] Its scientific name – *Agrilus biguttatus* – might sound like the name of a villain from Monty Python's *The Life of Brian*, but sadly the presence of this creature is no laughing matter.

The beetle is native to the UK, and until recently its recorded numbers seemed quite low, to the point where it was even considered a vulnerable species. Like AOD it is found in central and southern England and Wales.

The oak jewel beetle is partial to an old oak tree. The female lays its eggs on the tree and the resulting larvae tunnel inside, zig-zagging their way through to the yummy inner bark and cambium (tissue between the inner bark and the sap wood). It leaves D-shaped holes where the adults have left the tree. These D-shapes are the smoking gun. They often seem to be present in cases of AOD; the fingerprints that we keep on finding at the scene of the crime. And so this beetle has been implicated in spreading AOD.

But could it be that this creature has been wrongly accused?

Perhaps our beetle is just always in the wrong place at the wrong time, rather than a villain? Maybe. Other organisms that are harder to see but that also turn out to be particularly prevalent at the scene of the crime are certain bacteria including one called *Brenneria goodwinii*. On close inspection it might be the interaction between all these elements that brings on AOD; for example, it might be that the beetle is helping to spread the bacteria and/or it could be that these various factors are aggravating existing problems. Rebecca says, 'The co-existence of *Agrilus biguttatus* and the bacteria species could be incidental, they may be taking advantage of stressed trees and inadvertently creating a positive feedback loop by causing increased stress within the tree.' Relatedly, early studies indicate that our oak jewel beetle might be drawn by the smell of certain bacteria towards trees that are already stressed.[5] That would make some sense in that the soft wood of rot holes might well make for easy burrowing for the larvae.

What is clear is that the pile-up of pressures our oaks face is undoubtedly making them more susceptible to AOD, along with other threats. Climate change appears to be an accomplice, for example, exacerbating drought in dry summers. Waterlogging and high pollution levels are all potential suspects too. So, while a healthy tree can withstand some challenges and fight off pests and disease, particularly those it has evolved with – the native oak jewel beetle included – what with one thing on top of another it all gets too much. So our trees, well, they decline, they start to die back.

Rebecca tells us that the role of the oak jewel beetle 'is inconclusive' and we can't say AOD is definitely an infectious disease although, until we understand more, we should practise good biosecurity around trees with AOD symptoms.

All is not without hope here. We seem to have some examples of recovery and current evidence supports reducing stress and creating optimal conditions for the tree species as the best option for management.

So whether our little jewel beetle has a leading role to play or not, it doesn't feel at all right to lay all of the blame for AOD at its D-shaped door. We need to look at the bigger picture.[6]

THE OAK PROCESSIONARY MOTH

In 2006 a new creature, a kind of moth, was spotted on UK shores. This moth develops from a furry, grey caterpillar with a black stripe. These caterpillars are hairy, kind of sweet-and-cuddly looking. They make nests of silken webbing in oak trees. They hang out in gangs and the gang members do this odd little thing in that they follow each other around the tree. Cute, eh?

Not so cute if you touch one. Brush against one of these creatures and you can get a rash – a small number of people react strongly.[7] This is the oak processionary moth (*Thaumetopoea processionea*). Unlike the oak jewel beetle, it isn't native to the UK. It is native to central and southern Europe.[8]

The oak processionary moth (OPM) is causing problems in parts of southern England, partly because it can strip oak trees of their leaves. However, the biggest issue in this case is less about the direct risks to oak trees and more about the impacts of the caterpillar on humans. Interesting as these creatures are, it isn't a good idea to hang out with them. The knock-on risk is that this beastie could put people off oaks in some places, which in turn means they will not nurture them or plant them. 'Spot it, avoid it, report it' is Forest Research's advice, should you come across the caterpillars or their great, silken webs.[9]

None of this is the moth's fault, of course, it didn't ask to come here. And in fact, it was probably brought in on imported oak trees. This is a familiar story. Here in the UK, potential predators might need time to catch up and realise the caterpillars could be food. As noted earlier, while evolution can do incredible things, it needs time and space to do them.

Despite this, the UK has continued to import oak trees: 1.1 million oaks were imported between 2013 and 2015. Sometimes OPM has been imported at the same time, inadvertently but worryingly. A nest was even discovered at the Chelsea Flower Show in 2016.[10] When it comes to OPM we are now spending millions in research and practical action to deal with this problem, resources that could otherwise be spent on so many positive conservation actions. The whole thing is a great shame.

The oak processionary moth is just one of many imports affecting our trees. Since the year 2000, we in the UK have imported 16 new tree pests and diseases. At one point we were bringing in a new tree problem about every 1.3–1.4 years.[11]

Some pests come in on plants grown in other countries. Others take more adventurous journeys, travelling on unorthodox forms of transport such as packaging materials, car tyres and our shoes. In some cases we import the same disease over and over again.

If only we'd had a bit of warning about the devastating effects of importing tree pests and diseases we could have avoided problems like this, couldn't we? A warning, say, like the fateful import of disease-infested elm logs, which led to the loss of tens of millions of elm trees, many of them back in the 1970s.[12]

Or a warning like ash dieback, which has been rampaging through our woods since around 2012. In most parts of the UK we don't have to travel far to see ash trees dying under our noses. The economic cost alone is – again – immense. We think ash dieback alone could kill 80% of UK ash trees.[13] And there is a higher cost still. In the case of the ash we know of more than 900 species – plants, animals and fungi – that use the tree for their food, homes and security.[14] Who knows how many more are at risk? The most vulnerable of these are, of course, the specialists, those that can't live on anything but the one species. Threaten the oak and we weaken at least 2,300 species, 300 of which we know need the oak tree for their very survival.

How many warnings do we need to stop – inadvertently – bringing in plant diseases to the UK?

OTHER PLANTS

It's not only pests and diseases that are a problem, invasive plants can devastate whole swathes of delicate nature too. The plant collectors who brought the common rhododendron (*Rhododendron ponticum*) to the UK were probably thrilled with themselves back around 1763. What an interesting ornamental plant it made with its luscious purple-pink flowers. How well it took to our climate.

What an enormous and expensive problem it has now become. It has invaded sensitive areas including around 40% of our rainforest area. Out of its native habitat this type of rhodi becomes a beast, stealing the light, crowding out other plants. We can remove this plant, and there is some great work going on to do so, but at a huge cost. More millions are needed to solve this human-made problem.

Then there are the non-native tree plantations in our most sensitive, ancient woodlands. These are the old woodlands that have persisted since 1600 in England, Wales and Northern Ireland, and 1750 in Scotland. (The figures relate to the times maps started to be become fairly accurate – before this time it is unlikely the woods were planted by people because that sort of thing didn't really happen then. If woods were on these maps, then they are very likely to be natural in origin and have some form of continuity for many hundreds of years prior to that.) These woodland treasures are relatively undisturbed by we people, and so have built up complex, intricate, special communities not found elsewhere; communities of plants, fungi, animals and microorganisms over hundreds or, more often, thousands of years.

Plantations of non-native trees in ancient woods seemed like a good idea back in the early twentieth century following the world wars. We had felled great swathes of trees to support war efforts. Now we needed to start rebuilding. We needed more timber. We needed it quickly. So, at the time, it seemed to make sense to replace slow-growing native trees – oaks and all – with fast-growing conifers, but now we know we need to remove the latter to restore those sensitive communities. The Woodland Trust says, 'Almost 40% of the UK's ancient woodland has been replanted with dense non-native trees.'[15] This means we have another, immense task on our hands – we need to restore our ancient woodlands. This involves removing most of the non-native trees and doing it gradually, gently letting light into the wood and allowing native plants to re-find their feet. A delicate touch is needed; it has to be done slowly. Go too quickly and the gaps we create fill with plants such as bracken, which can take over and crowd out sensitive species.

There is of course a place today for fast-growing timber crops on UK shores – we need sustainable supplies of wood. The key thing is to grow them in the right places, and ancient woodlands aren't the right places.

The work goes on.

DASTARDLY DEVELOPMENT

A photograph went viral on social media in autumn 2020. It showed a woman in despair, head in her hands, crouching on the great, flat stump of a 300-year-old oak tree. The great trunk of the tree stretches out behind her. It was The Hunningham Oak near Leamington in Warwickshire. Newspapers reported it was killed to make way for HS2, a new, high-speed railway.[16]

The iconic picture captured the dejection of environmental campaigners across the UK. They were trying to prevent the destruction caused by the development of HS2, which has been tearing up veteran trees and ancient woodland – oaks and all – across the land.

You might say public transport is a good thing, right? Yes, usually. Environmentalists are, of course, generally all for travel that causes the least pollution possible. However, in places, HS2 has been inexplicably insensitive to our precious, often irreplaceable nature. It was widely reported that The Hunningham Oak was not in the path of the railway track but rather a service road. Its mistake was to spend 300 years growing in the way of some lines drawn up by the person planning that part of the railway. This doesn't make sense. Why wouldn't you plan a side road that avoided destroying a living monument that was, growing, shading and giving life and home to other species probably even before Catherine the Great ruled Russia?[17]

Why wouldn't we plan to support nature and the historic culture that comes with it? At the same time as Sir David Attenborough warns 'the world is in trouble'[18] and King Charles III calls for swift action on climate change, why would you design a railway that could smash through 108 irreplaceable ancient woodlands? And which destroys special, veteran trees?

Campaigners saved some of the ancient woods on the HS2 route. Others have been lost for ever. Broadwells Wood in Warwickshire was not only destroyed but, to add insult to injury, this was done in the spring, just as nature was bursting back into life. The Woodland Trust says: 'HS2 Ltd has not always followed best practice when removing these woods and has broken its own assurances that it would try to minimise the damage done. For example, ancient woodland in Warwickshire was removed in April 2020, against all professional guidance and HS2's own commitments to do the work in late autumn when the woods would be dormant.'[19]

At the time of writing, the government has announced a halt on parts of the HS2 plan, albeit largely for economic reasons. Meanwhile more than 1,000 irreplaceable ancient woods, nature's own cathedrals, are at risk from developments relating to roads, rail, housing, agriculture, electricity, gas, water and telecommunications. Our irreplaceable ancient forests have faced damage from everything from paintball centres to nudist camps. I have nothing against any of these things, except when they bring about the destruction of our natural heritage.

Campaigners are now fighting not only outright destruction but also 'helpful' ideas such as so-called 'habitat translocation' from those determined on destruction, as if to say, 'Don't worry, we'll pick up all the soil and just pop it down somewhere else'. This is a bit like proposing to smash up the *Mona Lisa* or Westminster Cathedral, say, into billions of parts and then stick it back together. Except smashing up an ancient woodland is worse; a talented team might be able to recreate an artwork or a building. 'Habitat translocation' is misleading; an ancient wood is a unique habitat, it is about geology, soils, aspect, microclimate, and centuries of history.

How could you translocate such a habitat and rebuild it without tragic loss? Think of those intricate underground fungal networks we visited in Chapter 6. And given that 75% of woodland carbon is in the soil, the impacts go on and on.[20]

Damaging our own environment is not new, of course. The poet John Clare captured the tragedy of the Enclosure Acts, which changed the shape of rural Britain, tearing up oaks, the countryside and the livings of many poor people tragic for both nature and people.[21]

> By Langley bush I roam but the bush hath left its hill
> On cowper green I stray – tis a desert strange and chill
> And spreading lea close oak, ere decay had penned its will
> To the axe of the spoiler and self interest fell a prey
> And cross berry way and old round oaks narrow lane
> With its hollow trees like pulpits I shall never see again
> Inclosure like a Buonaparte let not a thing remain
> It levelled every bush and tree and levelled every hill
> And hung the moles for traitors – though the brook is running still
> It runs a naked brook, cold and chill
>
> JOHN CLARE, EXTRACT FROM 'REMEMBRANCES'

Destruction of ancient trees and woods shows a turning from the lessons of history, a turning of the head when convenient. We need to understand the real cost of destruction and the cost of repairing it, if indeed we can. And we need to include the many hidden costs such as the time and resources campaigners and charities have to give to protest; to stay on top of planning; to prevent absurdities that damage our trees, our environment and, when you take it to its logical conclusion, our own future.

CHOPPING THAT NEEDS STOPPING

As if insensitive development isn't enough, we have also developed a related societal habit of chopping up our landscape. The modern way is to divide our world into different 'service areas'. In one place we have a town. In another we have a road. Somewhere else, an out-of-town centre. Here we have an industrial centre. There, an intensive farm. If we're not careful – and we have a tendency not to be careful at a collective level – nature ends up trapped on little islands surrounded by a sea of impenetrable barriers it cannot cross. Its own parts aren't connected to each other. The wild things can't mingle with others of their species. And there is no way for them to deal with threats such as flood, fire, climate change or habitat loss in these contexts – there is literally no escape.

So let's think about links. Let's reconnect our landscape along the lines suggested by Louise Hackett in Chapter 4. Nature is so much more resilient to all the threats it faces if it can spread out, and move around to the extent that it is able to move. A bird that loses its home, a woodpecker, say, is relatively mobile; it can move from one tree to another, perhaps even fly to a new wood as long as other trees aren't too far away. A bat might be OK if it has a green corridor to follow, a healthy hedgerow, say. A ladybird can fly from one place to another; a longhorn beetle can travel a little. A hedgehog might be more easily scuppered. It might find its own hedgie highways blocked by a fenced-in garden, and then it might have to take its chances crossing a road and we all know how that ends. A rare ancient-oak-dependent beetle has very few options, in fact none at all if there isn't another aged oak very near the one it has just lost.

Hope comes with some great initiatives: The Wildlife Trusts' Living Landscapes; the Woodland Trust's Treescapes;

the work of the Rainforest Alliance; the wilding of the Knepp Estate and others. Power to their elbows.

CLIMATE CRISIS + NATURE CRISIS = HUMAN CRISIS

Pepe and I are out for an early morning walk. The air is full of sweet, sweet song. Nearly every one of the big oaks on Hopyard Lane in our village seems to have its own robin up there in its own world, whistling away. Back in the garden there are red campions and feverfew in flower and marigolds in bud. It feels like spring and it is beautiful.

It is also a bit worrying. It is New Year's Day 2022, the warmest New Year on record in the UK. The Met Office wrote, 'New Years Day has also seen very mild temperatures with 16.3 C recorded at St James Park. This beats the previous record for New Years Day of 15.6 C at Bude (Cornwall) in 1916 setting another provisional record'. Over this New Year records for high temperatures were set across the UK.[22]

Nature's Calendar,[23] a citizen science initiative, shows us that spring now comes to the UK – on average – more than eight days earlier than before (back in the years 1891–1947).[24] It would be forgivable to think an early spring sounds rather pleasant. Here in the UK we've had iffy weather for several millennia, to the point where it is a national obsession. If we get more warmth, more sun, could that on some level be a good thing? Our oak trees enjoy the sunlight, don't they?

The problem is the speed of change and the chaotic nature of it. In wildlife terms, things are already getting out of sync. Different species respond to temperature changes in different ways and to different degrees. This can disrupt the dependencies in the world of the oak and beyond. Changes to the timing of tree leafing affects the timing and abundance of tree caterpillars, which has knock-on effects for birds. And that can mean a problem for blue tits, great tits and pied flycatchers, where chicks hatch too late to take full advantage of peak caterpillar numbers. You have to feel for them all, don't you, especially the pied flycatchers, who have flown all the way from Africa for the UK spring and arrive in great need of a good meal.[25]

So climate change means that bird populations can crash more easily. That is very sad for the birds. It isn't great news for the oak either. It means that birds aren't performing their usual role of keeping the caterpillars in check. The leaf munching is less controlled, which in turn means the oak has to make more effort to produce new leaves. Meanwhile, of course, all the other wildlife that depends on the oak's leaves also has a tougher time. And there are the other consequences of climate change: more droughts, more floods, and even more fire, potentially. The more change, the more unpredictability, and the harder it is for our wildlife to cope.

We are piling up the challenges for evolution itself.

Big, isn't it? Can we really do something about it? Well yes, and we must. This isn't just because it is nice to have oaks and blue tits and moths and all the other joys of nature. We are part of nature too. Looking more globally, we need a world in which huge swathes of humans aren't displaced because their own nature is depleted, flooded or dried up.

SNEAKY, SEEPING THREAT
Next up in our lineup of baddies is a silent killer that sneaks up and seeps up and invades and pervades the wider world of the oak: pollution. We touched on this earlier when we looked at rainforests, lichens and mosses. And, while we talk a lot about carbon emissions, there are other emissions that we need to tackle too. For example, as the Woodland Trust says, 'Excess nitrogen, including ammonia, has many

impacts on the natural world. Our State of the UK's Woods and Trees report identifies nitrogen air pollution as one of the most widespread and significant threats to woodland ecosystems in the UK.'[26] The Plantlife report 'We Need to Talk About Nitrogen' highlights the extent of this problem: 'Global pools of reactive nitrogen have been building in the atmosphere, soils and waters from burning of fossil fuels and intensive farming. This excess of reactive nitrogen is now being deposited throughout the biosphere, significantly impacting our most precious semi-natural habitats, changing their plant communities and the very functions these ecosystems provide.'[27] In many wild, natural places, reactive nitrogen tends to be scarce but, says the report, 'Once emitted to the atmosphere, primarily from human activities, reactive nitrogen may be deposited to soils and vegetation, where it can acidify soil and over-fertilise sensitive ecosystems.' As well as these impacts on ecosystems, reactive nitrogen contributes to human health impacts of poor air quality that costs billions each year.

What can we do about it? In trees and woods terms we can help by buffering woods and trees from the worst impacts – new trees can help. But the problem is wider than just for woods and trees, so for that we need action.

In its report, Plantlife and its partners said, 'Links need to be strengthened between related policy areas such as agriculture, water quality, energy, transport, climate change and public health.' We need to pay more attention to the interdependencies in life, something the oak and its world can help to teach us.

We have seen decision makers turn around the effects of pollution before, with the Clean Air Acts. Come on, politicians, we can do this.

DEER, OH DEAR

What's not to love about Bambi? Deer are great, beautiful, graceful. We have native UK deer, and sustainable deer populations browsing in woods are all part of the balanced ecosystem.

The problem occurs when things get out of balance with their environment. Too many deer, all munching away, prevents new regeneration, so young trees can't get established; ultimately it harms the woodland structure and reduces the biodiversity in woods.

Red deer and roe deer are native to the UK. Fallow, sika, muntjac and Chinese water deer have been brought here and settled. Populations of wild deer have been increasing rapidly in the last few decades, and may now be higher than at any time in the last 1,000 years. Deer are an important part of the UK's woodland ecology and have a key role to play in balanced woodland and wood-pasture ecosystems. The challenge is when that balance is lost, and in many parts of the UK deer have reached such numbers that they threaten the very habitats they and other species need.[28]

In the natural course of things we would have top predators; lynx and wolves used to live in the UK and would have kept browsing mammals in check. Could we bring them back? Should we? The problem is that we live on crowded islands and these animals need big territories. Personally I can conceive of the shy lynx living in the wilder parts of the UK. It is controversial, though. Even more contentious are wolves, about which the old fears and fables die hard. In his podcast *Trees a Crowd*, the happily named David Oakes talks about the issues related to higher-than-natural deer populations, including the number of car accidents involving deer.[29] According to the RSPCA, 700 people are injured and approximately 10–20 killed each year because of vehicle collisions involving deer, either through direct collisions or as a result of swerving to avoid deer.[30] David asks which would cause more damage: reintroduced wolves or the excessive numbers of deer, saying, 'Wolves seem a whole lot more natural.' It is a good point and it begs the question: should we really be afraid of the wolf?

The idea of reintroducing wolves makes an interesting discussion, and any decisions would need to be based on sound evidence. For now, though, what can we do? Well, if you eat meat, have you tried some local wild venison? It is healthy, in abundance and more widely available than it used to be. It makes sense.

A GNAWING PROBLEM

It isn't only deer that damage trees; other mammals such as squirrels sometimes have a go, stripping away at the bark with their sharp little teeth. This pastime can be damaging indeed to a tree, especially where the squirrel circles the whole trunk, cutting off the vessels that carry food, water and nutrients around the tree. In a human it would be like cutting off our blood supply.

The imported, invasive grey squirrels come in for a lot of criticism. In most areas of the UK, they have famously taken over from the native reds, out-competed them, spread squirrel pox virus and generally earned themselves a bit of a bad reputation. It isn't their fault. Like oak processionary moths, thanks to human actions they have ended up in the wrong place. A long time ago they were literally picked up and carried – presumably terrified – to a foreign country, one that hadn't grown up with them around. They are the unfortunate imported. And now they are the bad guys – numerous bad guys. So, anyway, we have an extra, gnawing problem. Their presence knocked a chunk of the ecosystem sideways and out of balance, adding another wobble in the wibbly wobbly tower of Jenga that is our natural world.

Why squirrels chew trees in this way, and to such an extent, we aren't quite sure. If we only knew, we might find ways to counter it and avoid a heap of problems. In some areas the problem is so pronounced that land managers can be put off planting trees. And there is a huge impact on the timber industry. Squirrels cost the industry millions per year.[31] The Woodland Trust's Dr Chris Nichols shared several theories with me: 'The squirrels might have a seasonal calcium deficiency – there is a lot of calcium in the phloem [phloem is the tissue that transports food made by leaves]. It could just be that they are after the sugar in the phloem as it comes down the tree. It might be about self-medication against parasites. It might even be a be a show of fitness, basically young male squirrels showing off with a bit of sexy bark-stripping to impress the lady squirrels. They might do it to get the water. Or they might do it for the calories. What we do know is that it is a problem when population densities get high in areas where we have a lot of grey squirrels.'

Squirrels are of course planters of oak trees (more on this soon) and we can wholeheartedly embrace our red squirrels. As always, in a natural world that is so affected by human behaviour, it is back to that balance. The bottom line should always be that we need more joined up landscape-level coordination by we people.

TREADING AND TRAMPLING

Thursday 15 October 1987 was a night of great destruction. Wind speeds of up to 115 miles per hour (up to 185 kilometres per hour) battered buildings, lifted rooftops and took down around 15 million trees in the UK.[32] One of those trees was The Turner Oak in Kew Gardens. This aged tree beauty, a cross between the pedunculate oak and the holm oak, was upended, turned on its side.

The poor staff at Kew's arboretum, led by Tony Kirkham, arboriculturist and co-author of *Remarkable Trees*, popped it back in the ground without much hope for its survival. They had so much damage to deal with all across Kew Gardens that they didn't have much time to give to this tree. Three years passed and Tony returned for a better look at The Turner Oak. It was in fine form. In fact, surprisingly, it was doing much better than before the storm. It was rejuvenated. It was positively flourishing.

'The Turner Oak was planted in I think around 1790 and I photographed it just before the storm,' Tony told me. 'When I look at those photographs now it was a much smaller tree, and since 1987, only 35 years ago, it's put on a third of its growth.'

After some scratching of scientific heads, Tony and team realised that, before the storm, the tree had been stressed. All those people passing and sheltering beneath it had squashed and compacted its root plate and this had done it no good. The storm and the uprooting had shaken things up. It had also, in many ways, done the world a favour: Tony and team had learned that compaction of soil around trees is a serious issue.

'Because everybody is walking on it, when it's wet, when there are wormcasts, they are all adding to compaction and then we're using heavy grass-cutting equipment and I just put two and two together and I thought, all our trees are suffering from compaction,' he told me. In other words, oaks need air, and they need it down below.

After seeing the significance of trampling around trees, Tony and team spread the word and tried out various devices to help deal with the problem, eventually settling on a machine to help out trees in similar predicaments. It was originally designed to squirt air at clogged-up expansion parts on bridges. These bridges are designed to expand in the heat, so you need to keep their spaces clear. But, it turns out, they have other uses, such as injecting air into the ground.

BACK IN SHERWOOD FOREST, a few days after my chat with Tony, I am with some ancient tree experts, including ecologist Jim Mulholland, whom we met earlier. He talks about a root survey on The Major Oak. The survey team used a kind of radar that could pick up roots two centimetres in diameter and more. Jim invites us to walk out to how far we think the roots might go. We spread out far beyond the 'drip zone' – where the leaves stop – and keep going. I have a feeling that this will be further than I think but don't really know how much further. Luckily I am walking behind a man called Tom Reed, who knows his ancient trees. We walk and walk until we reach about 40 metres (about 130 feet). This is where the roots go to. By that I mean the roots we have measured. So far.

I looked up something we can relate to online. An Olympic-sized swimming pool is 50 metres (about 164 feet). The key thing is that tree roots spread further than you imagine. Much further. It goes to show how we need to respect space around trees, especially our most precious, ancient trees. These trees need protection from trampling, as well as needing space and also light. The more we can learn to read the signals, the better off we'll all be. As Tony says, 'Trees are like people. They are moody, they stress, but they're beautiful when they're happy.'

FALSE PROMISES

In the last few years we have all heard young people, perhaps most famously Greta Thunberg, express, often in exasperation, how they felt obliged to give up part of their future, how they have felt the need to go on school strikes and demonstrations to make governments take notice of the bigger threats to their – our – future.

At the 2021 Youth4Climate summit in Italy, Thunberg said, 'Build back better, blah blah blah. Green economy, blah blah blah. Net zero by 2050, blah blah blah. Climate neutral, blah blah blah. This is all we hear from our so-called leaders. Words that sound great, but so far have led to no action.'[33]

The climate crisis we face is intrinsically bound up with the nature crisis. Sometimes we see the right signs from governments, but sometimes we don't. Recent years have increased protection for ancient woods; however they – despite being irreplaceable – are still being destroyed. Meanwhile ancient trees have little serious legal protection.

On planting we have seen government commitments fall short year after year. In the 'State of the UK's Woods and Trees' report, the Woodland Trust says, 'Tree cover in the UK is increasing, but nowhere near fast enough, particularly native tree cover. Over the last five years, the rate of woodland expansion has been on average just under 10,000ha per year – with 45% comprising broadleaved trees. The UK is failing to reach anywhere close to the target of around 30,000ha per year that is estimated to be needed to reach net zero carbon emissions by 2050'.[34]

Hollow promises can be as seeping, creeping and insidious as the poisonous pollution threatening our nature, because they keep us quiet. They silence us. People think we have won the argument, reason has prevailed and they move on with their busy lives.

Governments have difficult balancing acts to perform, that is for sure. When it comes to nature, we have to be wary of compromises, though, because we need balance now more than ever. It is no exaggeration to say that our very survival depends on it. Health and happiness start with the very air we breathe and the soil beneath our feet. And we need people to understand this is just as much about us humans as it is about wild plants and animals. Nature is not something that is merely nice to have; it is a must, a need. In a webinar I attended, Dr Charlie Gardner, lecturer in conservation biology at the University of Kent, said memorably, 'Conservation is essential to safeguard human civilisation. So why do we market it as the altruistic quest to conserve other species?'

We need joined-up action to save our oak trees and our wider nature. We need government – national and local – to coordinate action that is beyond individual people and organisations. We shouldn't bash all politicians, we mustn't tar them all with the same brush. Absolutely not. Those who say, 'Oh politicians are all the same, all cynical,' aren't right in my view. I have met politicians who care deeply about people and planet but we still need to let them know how *much* we care and hold them to account.

We have stories from which to take inspiration. When the government tried to sell off our forests – those owned by the state – in 2010, we, the public, howled out in anger and we won that battle.[35] Earlier I touched on the Clean Air Acts, which made a difference to our environment and to human health. We have seen progress in both planting and protecting trees. Now we need to go further and to howl out again against all the creeping threats: climate change, pollution, false promises. This is the moment to do this. As Greta says, 'We can still do this... Change is not only possible but urgently necessary. This is not the time to give up.'

··· One in a Million: A Leafy Activity ···

So what can we do about threats to the oak and its world? So many things. Campaigning is vital and we'll find out more about adding our voices to campaigns shortly. Meanwhile, you know what? Just enjoying oak trees and letting them enter your world is important too. Here's a really simple autumn activity that we can all do, with kids or simply by ourselves. It is one way to get up close and personal with these amazing trees.

Visit an oak tree when lots of leaves are on the ground – and pick up some leaves.

How many different shapes can you find? Are the shapes regular? Does one side of the leaf mirror the other? Or are they crazily wiggly and irregular? How many different sizes can you find?

How many different colours can you find? Can you find green, gold, brown, bronze, yellow, auburn, orange? Can you find others?

How many words can you find to describe your leaves?

Do some of the leaves still have galls on? Do others have scars where the galls once sat? Do any have bite marks from creatures? What might have had a nibble?

If you can find another oak or two, do the same there.

Do the leaves from one tree have anything in common? Do they look different to those from another tree?

Pick up those that appeal to you and make a little collection, then spread them out on the ground. Alternatively arrange them into a collage: a face, an abstract picture or whatever takes your fancy.

Hopefully what will come across is the individuality of each tree and each leaf. And lots and lots of beauty. Each leaf is gorgeous in its own way. Like every tree, each leaf has its own character. Just sometimes we don't notice it because there are so many of them. If they were more rare perhaps we would. Meanwhile, each is special... they are a bit like us. Like you, each leaf is one in a million. OK, many millions...

GOING FURTHER
If you'd like to explore the leaves of more types of oak – beyond the UK natives – the website Leafy Place takes you through a great selection, which goes to show how much variety there is.[36]

There is also a lovely board game called Tree Bingo, created by the wonderful Tony Kirkham, which helps you learn about all kinds of trees.

. . . . . .

8

The Oak's Little Helpers. And Its Big Helpers.

THINK OF THE FIERCE ENERGY CONCENTRATED IN AN ACORN! YOU BURY IT IN THE GROUND, AND IT EXPLODES INTO AN OAK!
GEORGE BERNARD SHAW

LOVE

Love is a core component in looking after the oak and its world, motivating we humans to protect it; inspiring the practical things. Oh, and oaks need help from birds and mammals and other species like fungi that we met earlier. In this chapter we'll take a tour around some species that are doing their bit for oaks – including humans and the oak itself – and we will explore how love can help protect our oak trees.

SELF-HELP

First of all we should look at a spot of oak self-help. We have seen how the oak is a great survivor, and does pretty well at looking after itself in normal circumstances. It is doing a lot of the heavy lifting in the fightback against the threats we saw in Chapter 7, and it has some cunning strategies to outwit animals that want to eat it.

Earlier in the book, I touched on the way the oak can replenish its leaves with Lammas growth, even after being nibbled to near nakedness by caterpillars. Cleverer still – or at least I think so – are the oak's mast years. Every few years, oak trees, like beech trees, have a mast year when they produce a huge crop of seeds. In the oak's case these are the years when the ground is carpeted with acorns, such as it was in 2020.

One of the joys of escaping outside for a much-needed walk round our way in that Covid-plagued winter was you couldn't walk the country lanes without crunching on acorns. Meanwhile, conservation social media was awash with comments reporting on enormous acorn crops from all across the UK. It was a whopping great mast year. But what is going on with these mast years – how do they happen and why?

During a walk around a forest in 2022, I ask Lorienne Whittle, the former citizen science officer of Nature's Calendar, about this. Lorienne says scientists have identified a 'boom and bust' pattern in acorn production. In 'boom' years – such as 2020 – oak trees produce more acorns than the animals that feast upon them can possibly eat. These are followed by 'bust' years, when seed production is lower and trees instead focus on growth. For example, 2021 saw particularly low numbers of acorns.[1] The theory is that animals will struggle more for food in non-mast years, so this helps keep their numbers in check; a natural kind of rebalancing. When it comes to a mast year there is plenty to go around; more acorns will escape hungry mouths and grow up into trees. One of the strange things about mast years is that trees somehow coordinate across a vast area. Beech trees, for example, have been known to produce super-abundant crops at the same time across the whole of northern Europe. We don't yet fully know how, Lorienne says. 'Weather seems to play a key role because, for a healthy crop, the trees need the right combination of temperature and rainfall in the spring.' Oaks produce a similar number of flowers each spring, so the size of autumn's acorn crop depends on how many of these flowers are able to mature into ripe acorns.

OAK SEX

The way oaks make babies also has a touch of resilience brilliance to it.

We saw earlier that oaks, like many trees, are arguably both male and female and arguably, in a way, something else. So how, you might well ask, do they make baby oaks? How do our trees make whoopee? Let's go there: oak sex.

Wind is important. The breeze plays Cupid in that it carries pollen from the male flowers to the female. And, because the flowers are often so high above the ground, the pollen can travel a long way. As a result, the acorns on one tree can have numerous different fathers. Imagine that: thousands of acorns, numerous beautiful little nutty acorns, lots of different dads. This is a blended family par excellence.

What is important here is less about the joys of tree free love, and more about genetic diversity. Acorns have varying characteristics. Differing the parentage increases those modifications further. These variations give the oaks flexibility to adjust to different conditions. So oak reproduction is a clever thing in itself, allowing the tree to adapt to change.

The challenge the oak has now is that big changes – caused mainly by human activity – are so quick and on such a big scale. The oak, like the rest of nature, is having to adjust at record-breaking speed. Time is the key. But some help, at least, is at hand.

ANIMAL AID

The oak has a number of little helpers in its ongoing quest to survive and thrive. The jay and the squirrel, for example, are famous as oak planters extraordinaire. And there are a few more characters that seem to be lending a little hand in the world of the oak. These range from a heart-shaped surprise to a prickly customer to something very close to home. Let's start with the jay.

The jay

It is autumn. A low sun is slicing through the oaks in the grounds of Belton House, a National Trust site in Lincolnshire. Orange-auburn leaves are waving back and forth, creating short-lived gaps. They let in the light and then shut it off again. Each time the sun pierces through, a sharp twinkle catches our eyes. The whole canopy is winking and sparkling. Stars in the daytime.

My walking companion – also my line manager at work, Karl Mitchell – and I are blinking in the beauty of the scene when onto the stage screams a character. It brings fresh colour, light and joy. It is the jay.

Karl, who is better at birds than I am, identifies the newcomer. I ask how he knows; what alerted him to the jayness of our bird. 'It was the flash of the pink,' he says, referring to the bird's dusky blush of a body; a soft, brownish pink, the colour of fading tea roses. This is set off with a burst of white bottom and it all clashes beautifully with striking electric blue go-faster-type stripes on the jay's wings. This bird is a quirky beauty of contrasts. Get up close to a jay – or take a look at its picture online – and you will see a slightly angry-looking face topped off with a dapper, streaked hairdo and a black handlebar moustache. The jay is avian steampunk.

Our bird interrupts us by crying out its own name. 'Jay, Jay, Jay!' it shouts in a strange, slashing voice, as if someone is tearing bits of old cotton sheet to make rags.

In a blog post for the RSPB, the conservationist Jamie Wyver writes, 'As they've been part of our countryside for a long time jays have a variety of old, local and alternative names. A few describe their screeching call: devil scritch, scold and the Gaelic "schreachag choille" meaning "screamer of the woods". There's also 'oak jackdaw', which refers to the jay's penchant for acorns.'[2]

The jay's scientific name is *Garrulus glandarius*, meaning roughly, and gorgeously, 'chatty acorn eater'. And, true enough, it loves to dine out on a tasty, filling acorn, or two... or two thousand. Jays can collect and stash 2,000 acorns a year according to the British Trust for Ornithology (BTO), hiding them away under the leaf litter in the autumn.[3] These make up the jay's winter larder, a food supply to keep it going in the harsh times ahead.

Our jays are clever birds. They have a great memory, and can recall where most of their acorns are hidden. Any acorns that slip through the net and aren't dug up again and gobbled up could, one day, become like the amazing Ancient Burry Oak on Belton House's estate.[4] Just give them a few hundred years and a bit of luck.

When I say a bit of luck I mean – as always – a lot of luck, because, despite its thousands of seeds, a mature tree can expect very few of its acorns to become trees. And to reach a really great age, that tree must somehow outwit disease, damage, development and pollution.

Jays don't only plant oaks, they also carry some of them to superior accommodation. That's if nibbling animals don't get them first. In many cases the jays carry the acorns out from the dark part of the forest and bury them in a lighter, brighter spot where their chances of surviving are improved. It is as if some fortunate seeds have their own little bird airline with a special acorn-sized beak seat.

Even better, as RSPB CEO Beccy Speight says, 'The jay can help oaks travel uphill. Acorns are heavy; they don't fly like birch seeds or take to the breeze the way ash keys do or catch the wind like the twirling moustaches that are maple seeds.'

In this way, jays help a woodland to shift and expand, to move around as long as it has space to do so.

Again I am reminded of the witches' prophecy in *Macbeth*, of a wood marching on Dunsinane Hill. The advance would be a slow one, of course, taking place in tree time, not really quick enough to catch the human eye. And certainly not quick enough to determine the outcome of a battle.

As this book is driving towards the importance of love and its role in protecting the world of the oak, I should mention jays seem to have a soft, romantic side. They mate for life. And they have been seen to give their partners little love tokens. Not just any gifts though; proper, thought-through gifts. They seem to anticipate their partners' needs. Experiments describe a scenario in which male jays watched their mates eat their fill of either moth or mealworm larvae. They were then offered a meal choice – moth or mealworm larvae. The male jays tended to gift their partner with the food Mrs Jay had not yet eaten.[5] They seem to have worked out what their lover needed and wanted. Perhaps we could learn a lesson or two in great gift-giving from our jay friends.

Jays are excellent mimics. For some, currently unknown, reason they seem to enjoy imitating tawny owls and buzzards.[6] I like to think they enjoy having a comedic pop at those big, iconic celebrity birds. A more plausible theory is that the more tweets and sounds they can make, the more attractive they are to potential mates. They are a bit like the fun exhibitionist in your teenage gang, always drawing the attention with their impressions of stars and celebrities. This talent for impersonation might well be about extra added powers of attraction.

Before we get too sparkly eyed about jays, though, I should probably also mention here that they sometimes eat other birds' babies.[7] But hey, no one's perfect. I hope you, like me, can love the jay for the character it is.

The squirrel
Another creature that famously does its bit to get oaks in the ground, albeit not entirely without controversy – as we saw in Chapter 7 it can also nibble our trees to death – is that bouncing bundle of woodland whizziness, the squirrel.

Like the jay, the squirrel famously curates its own stores of acorns and, again, some of the lucky seeds will escape the eating to sprout.

Our native red squirrels, which go by the scientific name of *Sciurus vulgaris* – 'vulgaris' might sound a bit rude but means common or ordinary here – currently survive only in pockets. They can be found in – for example – parts of North Wales and Scotland, the Isle of Wight, Formby (near Liverpool) and in the Lake District. These groups, small as they may be in places, give hope, and we can give them a helping hand by providing the habitat they need to thrive.

In this chain of oak-related life there is an unlikely oak hero who I'd like to give a nod to here if only for surprise value: the pine marten. The heart-faced, cream-bibbed, bushy-tailed pine marten looks like a long, slim cat and it appears to be helping red squirrels. The pine marten was once hunted almost to extinction, and studies reveal that where it has reappeared, red squirrels show signs of revival. It seems that pine martens tend to catch the grey squirrels but less so the reds, which have co-evolved with the pine martens. The grey squirrels are 'predator naïve' in this context. The reds weigh less, too, and can escape to branches and twigs too fine to take the pine martens' weight. Reducing the number of grey squirrels and increasing the number of reds might – in theory at least – help the oak, as the greys cause more damage. It will be fascinating to see how this story develops.

More little helpers

While the jay and the squirrel are perhaps the most well-known helpers of oaks in that they bury acorns, what I hope we have seen throughout this book is the importance of the way nature links up. In previous chapters we saw how vital fungi are to our oaks; how key the songbirds are that keep the caterpillars of the moths in check. Even humble thorny bushes – hawthorn, blackthorn and the winding briars of the dog rose, for example – can play a role by circling a baby oak tree, creating a kind of nursery that puts off nibbling animals and protects our seedling beauty from some of the ills of the world. This is about the whole lacy intricacy of the natural world.

And then there is humankind.

THE HUMAN TOUCH

We have seen how the oak has been doing OK for hundreds – well, thousands – of years, fighting its own battles with the support of furry friends, feathery friends, fungal friends and beyond. However, at key moments in history, it has also had a bit of help from we humans rolling up our sleeves and mucking in alongside the jays and the squirrels.

Cue the people planters.

In 1580 Queen Elizabeth I gave the royal thumbs-up to the creation of an oak plantation in Cranbourne Walk, in Windsor Great Park. It was, as Archie Miles says, 'the first record of a pure oak plantation'.[8] OK, so this wasn't necessarily driven by loving nature; there was this immense, intense demand for oak, not least for the tree-hungry ships, those 'wooden walls' needed in trade and in war. King Henry VIII's *Mary Rose*, described by Admiral Howard in 1513 as 'the noblest ship of sail of any great ship, at this hour, that I know in Christendom' needed 600 oaks in the making.[9] And this need was expanding. That pales into insignificance when we learn that in the mid eighteenth century, the building of the HMS *Victory*, Nelson's flagship at the Battle of Trafalgar, involved an astonishing 5,500 oaks, as well as other trees.[10]

War propelled planting. While seventeenth-century forest officers were encouraged to propagate oak and ash trees by throwing their seeds into scrub, in the eighteenth century the Royal Society for the Encouragement of the Arts offered prizes to prolific planters. The historian Simon Schama has noted that 'Acorn fever took hold. The great Dukes – Bedford above all – vied for planting out acre after acre with oak.'[11] Admiral Lord Collingwood, Nelson's second in command at the Battle of Trafalgar, is said to have carried a pocket full of acorns on walks in the countryside, surreptitiously scattering them in the parks of unsuspecting hosts; but, says Schama, 'The all-time champion was the lord-lieutenant of Cardiganshire, Colonel Thomas Johnes, who between 1795 and 1801 planted some 922,000 sturdy oaks.' The oak was 'England's tree of liberty' and sailors sang, 'Hearts of oak are our ships, jolly tars are our men.'[12]

And then, after all the oak mania, woods went out of fashion. Iron ships came in. Coal took over from charcoal. Timber became less important. Some forests were literally torn from the ground. They could be turned into fields and the people needed feeding. After all, woods and trees no longer held much value, right?

Then a great thing happened.

TO SEE HOW a great thing happened I visit Hainault Forest in Essex. It is winter. Most of the leaves have left the oaks and their fellow trees and are now busy mulching the ground. The exposed trees look like dancers in rehearsal; stripped of their showtime finery, reduced to the essentials, their dark, muscular limbs revealing every perfection and every flaw, a beauty less obvious than the festivals of spring, summer and autumn.

I am with Jonathan Jukes, who looks after this forest for the Woodland Trust. He talks about the disafforestation of this forest (it lost its old, legal status as a forest and was stripped of its ancient forest rights). That took place in 1851. The powers that be, with their Victorian machines, efficiently removed 100,000 trees, oaks and all, across 24,000 hectares (60,000 acres), within a few weeks.[13]

That's when the great thing happened: the bubbling up of a new movement, the conservation movement. The community, led by a chap called Edward North Buxton, MP, rose up on behalf of the trees. This is what we needed then and it's what we need now: MPs on the side of the trees. Anyway, people were outraged; letters were written to the *Times* and a few hundred acres of the forest were saved. This might have seemed like a small consolation prize at first. However, the rising had a wider impact; it captured imaginations; fired up hearts and minds and this helped save other woodlands, including the still-magnificent Epping Forest.

Hainault Forest – what was left of it – reopened in 1906 to great fanfare. Now, more than a century later, the local community – with the Woodland Trust and partners including the London Borough of Redbridge and Essex County Council – is once again protecting Hainault Forest, caring for the remnant of ancient forest and the veteran trees – it has some particularly special old hornbeams – within. They are expanding this special place, buying back bits of land alongside it and planting new trees – oaks and others – that will buffer and protect the ancient woodland while linking up the landscape.

Walking around this forest, I think about the naval surveyors of old, oak-dependent creatures that they were. They would have seen oaks like this through a different lens to ours. They would have eyed up each twist and turn of the great limbs. They would have seen strength and potency in the nooks and bends and they would have noticed the join where a branch met the trunk or where roots made just the right shape for a ship's knee. They knew how to go with the flow of the wood, its natural shape, harnessing the force and endurance of the oak for the ships on the seas.

Today human help for oaks is vital. And it is urgent. Oaks and other trees are vital in themselves. They are among our best warriors in the fight against climate change. They are stalwarts in staving off the collapse of nature and planting one creates a connection that will stay with us forever. It is an act that engages not only a spade, but also a heart. Woodland Trust CEO Darren Moorcroft said, 'It is an incredibly powerful act to stick a spade in the ground and plant a sapling. I think the connection that creates with that individual, the overall cause, the tackling of the climate and nature crisis, the doing something which will last for a long time. I think this is the most powerful way as a society we can engage new individuals, communities, in caring about the oak tree or the beech or the birch, opening up that broader world.'

And if anyone needed any more persuading that engaging hearts and minds is important, I'll have a go at explaining.

The UK government's climate change advisers have set a target of 17–19% woodland cover as part of the UK's plan to reach net zero carbon emissions by 2050.[14] This is no mean feat, even for an area as relatively small as the UK. The government has set a related target of planting 30,000 hectares (75,000 acres) of new woodland every year; roughly 30 million trees per year.[15] And to do that you have to find a place – the right place – to put them all on these, our already-crowded, islands. The scale of the challenge is highlighted by the failures of successive governments to meet their own targets. In 2020–2021 only around 14,000 hectares of new woodland was created in the UK.[16]

Why is it so urgent to find ways to reach these targets? The Intergovernmental Panel on Climate Change (IPCC) tells us, 'Temperature rise to date has already resulted in profound alterations to human and natural systems, including increases in droughts, floods, and some other types of extreme weather; sea level rise; and biodiversity loss – these changes are causing unprecedented risks to vulnerable persons and populations', and 'The most affected people live in low and middle income countries, some of which have experienced a decline in food security, which in turn is partly linked to rising migration and poverty.'[17] We need to keep precious trees in the ground, and we need to get more trees in there. The right trees in the right places help us fight flooding, overheating and pollution. They help us breathe, they provide homes for wildlife. They play a key role in enabling us to have healthy, stable communities. As Baroness Barbara Young said at the launch of the Young People's Forest in May 2019, 'If it wasn't for the tree you'd have to invent it.'

There is hope. Over the last few years we have seen much of the world wake up to the importance of getting trees into the ground. During the UK's general election in 2019, conservationists found ourselves in the unusual position of watching politicians of the main parties competing to make ever-greater pledges about tree numbers. To those who had been raising concerns about tree cover this was brilliant; the obvious proviso being that those political promises aren't always kept.

While statistics help us envisage the scale of the challenge, this isn't just a numbers game. Planting trees willy-nilly isn't the answer. We need to get the right tree in the right place. That means looking at the bigger picture again. We want that healthy landscape we looked at earlier in the book, that matrix of different kinds of nature in different places. The Woodland Trust's Dr Chris Nichols describes 'a diverse range of habitats at the landscape scale, with ecological resilience underpinned by ecological complexity'. Just as we need the right trees in the right places, we also need the right nature in the right place.

Caveats apart, this recent great tree awakening was a sign of progress. It was heightened further when the Covid-19 pandemic landed and people in the UK and across the world turned to nature, where they were able, as a source of solace and sanctuary, appreciating it more than ever. This appreciation and understanding is vital because it might be the magic ingredient we need to help us save our nature and thereby save ourselves. There is a wave of tree love going on across the UK and this is, of course, a great thing.

HOPE

On a cold Saturday in late November 2019, I found myself standing, spade in one hand, oak sapling in the other, on the site of a former coal mine. I was surrounded by hundreds of people – a great many of them teenagers and children. Together we were helping to create a brand new forest. This was to become the Young People's Forest, a brand new 400-acre forest in the heart of Derbyshire, in D. H. Lawrence country.

Game of Thrones star Bella Ramsey – aka Lyanna Mormont – themself a teenager at the time, was planting alongside us, mucking in with the crowd between requests for celebrity selfies. TV presenter and gardener Danny Clarke joined in, sharing tree-planting tips. One man was dressed as a tree and another, all in green – including his skin – and sporting a hairdo made of leaves and berries, was wandering around the field piping on a flute to keep us all going. Elyse White, one of the young people, then a teenager, leading the project, was busy giving interviews to the media. She spoke of her hopes for the future, for nature and for a forest that could be in the ground within three years.

There were pockets of ice on the ground; sun and mist were chasing each other about the landscape and were joined by the steaming breath of the hard-working planters. Together, that day, we dug into the hard earth of the site of the old mine and then gently placed our saplings, pencil thin as they were, spreading their roots into the unforgiving ground. We pressed and persuaded the soil back into position, bedding in our tiny trees: oaks, rowans, maples, birches, hazels and other native species. We now know that planting a mix of species will help keep this forest more resilient to tree disease.

We planted 10,000 trees that day and, as we planted them, people shared their reasons for coming. Some were there because they wanted to grow up with a forest. Others were planting for their children's future, the freshness of their air. The RAF cadets of 209 Squadron popped a fitting 209 trees in the ground. Members of the Sikh community bedded in 550 trees in honour of the 550th birth anniversary of Guru Nanak, the founder of the Sikh faith. Our own great 'planting out' Young People's Forest event was one of more than 100 'Big Climate Fightback' themed events. And more than a quarter of a million people pledged to plant a tree in this Year of Green Action. By way of an aside, that day was also notable for the amount of cake that got eaten. Now, there are very many reasons to get involved in nature conservation, you might wish to join to save the oak and the

2,300+ species it supports; you might choose to join because you love other trees, meadows, frogs, bees, butterflies. You might feel inspired by the fun, the fellowship, the friendships, the learning. You might join because you absolutely have to do something about the climate crisis, not to mention its twin, the nature crisis. But if all these reasons aren't quite enough, there is one more thing that might just win you over: cake. Cake and conservation seem to go hand in hand, cake being a kind of conservation fuel. It is made by staff, by volunteers and by well-wishers in great abundance. When it comes to cake, every year is a mast year.

Back in the burgeoning Young People's Forest we were standing back, pink cheeked and out of breath, stamping feet and flapping arms to keep our digits warm. And then came a bit of a downer: a couple of older men, locals, friendly chaps, flagged me down; they told me they loved this idea of a Young People's Forest but they wanted to manage my expectations; they wanted to warn me that the trees would likely be vandalised within a couple of weeks.

Three years on
Saturday 12 March 2022. Three years on, I am about to go back to the Young People's Forest to help get the last of the forest's grand total of 250,000 trees in the ground. The planting has continued over all this time but in fits and starts, as the young people and their community had to weave their activity around the challenges of the Covid pandemic. It is the final day of planting, which should be exciting but a cloud hangs over the morning. The news on the radio is awful, it is all about war. Ukraine has been invaded – the war is weeks in – and this as the world is on its knees, crawling through what we hope is the end of the worst of the Covid

pandemic that has ravaged society for the last two years. Social media is full of this conflict and other wars that have been raging around the planet for years. The stories are of people, animals and nature being smashed to bits by bombs from planes. Each of those planes is worth millions. Millions spent on destruction; resources that could have been used in the big fight to save our nature, the future of our children. I set off to the Young People's Forest, thinking glum thoughts.

On arrival I'm greeted by a group of smiling teenagers. Some are displaying the slogans #BigClimateFightback and #iWill on placards. Many of them are now adept planters and protectors of trees. They are showing members of the wider community – many of them much older people – how to plant a tree and they are talking about the wildlife that is beginning to populate the site: birds and beetles, frogs and hedgehogs. I'm thrilled to hear the butterflies are doing particularly well and remember the lovely butterfly recorder Ken Orpe, whom we met earlier, talking of his work supporting this site. This is a space where older people assume the best of young people; they listen with respect, learning from them.

I am also heartened to be told that, since we started planting three years ago, there has been very little vandalism on the site. That news and these people give me hope.

Sabaha Hussein, 19, one of the leading lights of the Young People's Forest and another of the project's leaders, says: 'The young people involved in this forest are like acorns, they are growing with the forest.'

Sabaha, Elyse, another founder member, Will, and a growing team of young people kept this new forest developing and growing through years of deep pandemic. They came out and tended the site whenever lockdown rules allowed.

Sabaha says, 'All through the pandemic everyone went through their own struggles, whether through loneliness or isolation. It made us realise how important nature is, not just because it is enjoyable but studies show how important it is for health; mental health as well as physical health: it affects you on a cellular level.'

By the end of the day the fledgling forest is fully planted and the world – at least the part we are standing in now – feels like a good place. A quarter of a million trees are now in that ground. We have a young forest that is also a Young People's Forest. We have a leafy beacon of hope and a community of people, both local and those who have bussed in for the day, who all care deeply about the world around them.

Sabaha says, 'On this project it has been so important to get the community involved. The experience of doing that in some way, you might look at trees differently. When I talk to people at community events they say, "I'm giving back to the world". This isn't an experience they are going to forget.'

Is planting the right idea?

But, hang on, should we actually be planting trees at all? There is an argument to say we shouldn't. Why not let nature do the work? Jays and squirrels are clearly well suited to the task and, if you let nature take its course, the trees will grow where they want to grow. Natural colonisation, natural regeneration makes a lot of sense. You can see it in action in places such as the Knepp Wildland Project near Horsham in West Sussex, a 3,500-acre estate of land – former farmland – which is the site of the rewilding project led by Charlie Burrell and Isabella Tree.[18] They are working to allow nature the freedom to be itself. The progress, along with the challenges faced along the way, have been captured in Isabella's book *Wilding*.

In an article for *Wood Wise* magazine, Charlie Burrell writes about trees getting established all by themselves and points out that they were designed that way; however: 'Our modern land management systems have just been preventing them from doing so. But in the Southern Block at Knepp, thorny scrub protects oaks planted by jays; sallow, birch and field maple blown by the wind; and crab apple and wild service dispersed by birds and small mammals. A landscape is emerging that looks like the ancient "forests" of the past – a kaleidoscope of habitats characterised by open-grown trees, groves and grazing lawns.'[19] We are back to that vision for the Woodland Trust's Treescapes, with oaks both in forests and – vitally too – in the open spaces where they might live to be ancients.

Biodiversity has rocketed, writes Charlie. Surveys show birds proliferating: nightingales, turtle doves and owls alongside other wildlife such as purple emperor butterflies. 'For the trees themselves, growing from local seed (possibly with regional adaptions), with their thorny nursery providing the micro-climate; and their roots in functioning soil, tapping into mycorrhizal fungi and bacteria, and benefitting from a wealth of nutrients; seems the likeliest scenario to boost immunity and resilience.'[20]

Allowing trees to colonise the landscape naturally is, without question, often the right approach, especially where nature is given enough space and has the right seed sources. However, sometimes, I would argue, we need to get out our spades and plant trees too. Joan Cottrell, of Forest Research, writes in *Wood Wise*: 'Where natural regeneration is a viable option, management is essential to promote the production of plentiful material on which natural selection can operate. This involves the promotion of seed production, the creation of space within woodlands for seedling establishment, and the control of herbivores so that enough of the established seedlings survive. Where natural regeneration is not a viable option or does not meet management objectives, planting stock will be required to achieve woodland expansion.'[21]

I think we need planting as well as natural regeneration. Right approach, right place; and some sites can benefit from both approaches, depending on the objectives of the landowner. Abi Bunker, director of conservation and external affairs at the Woodland Trust, says that native woodland expansion across the UK has an important role to play in the solution to climate change and enabling nature to recover, and that 'achieving this is going to require a spectrum of approaches for establishing trees'. On tree planting, she also notes how it 'engages people of all ages and social groups with the natural environment, often creating a lifelong connection with woods and trees. This action can and has been truly life changing for many people. Putting trees in the ground can also be an effective and efficient way of establishing woodland quickly and successfully, particularly where there is no nearby native tree or shrub seed source.'[22]

Meanwhile we need to get over other challenges. Land is expensive and has many roles to play. It has to provide us with food, living space, room for transport networks, the list goes on. Finding space to get the millions of the right trees – oaks and others – in the right place and relinking our landscape needs a concerted effort; great planning and cooperation are vital.

PROTECTING THE OAKS WE ALREADY HAVE

First do no harm. The age-old saying most often used in healthcare could well be applied to human activity in the natural world. It is a no-brainer that we should protect the trees we already have. Even a relatively young grown-up oak at, say, 200 years of age, cannot, of course, be replaced in our lifetime.

This makes it all the more astonishing that most of our trees, including many of our oldest trees, have no legal protection. As the Woodland Trust says, 'They're living legends. Pieces of history. Your heritage. They've witnessed centuries of change and been part of our landscape for generations. They're also vital havens for wildlife and important carbon stores, bringing health and wellbeing benefits for people and nature alike. But our oldest and most valuable trees are also vulnerable. Right now, most ancient trees have no great legal protection in the UK.' And ancient woodland, home to more wildlife than any other UK habitat, can still be destroyed because of the gaps in laws and planning rules.[23] These national treasures should be here to stay. They deserve the same sort of protection enjoyed by our beloved old buildings.

We have seen some progress in this area as well. In 2018, many years of campaigning paid off when the government strengthened protections for ancient woodlands and trees. This came about in the National Planning Policy Framework (NPPF). Now it is time to go further – we should protect every ancient tree, every ancient woodland because each is rare and precious.

I think this is where hearts and minds come into their own. The more we know about trees and woods, the more we tend to love them. I don't think anyone who has stood and looked in wonder at The Bowthorpe Oak or The Big Belly Oak, or felt the magic of a temperate rainforest, or held a wisp of lichen in their hand, or heard a woodpecker drumming its heart out can fail to fall a little in love with these trees and their world. To know one is to love one, and the more people get to know and understand a tree, a bird, a beetle or a fungus, the more they are likely to help protect it.

The more we share their tales and those of the species that live on and around them, the more our love of them will grow; of that I am certain. This is why we need to invite as many people as possible to step through the oaken doorway into a new realm.

There are human champions of nature from all walks of life; those who realise that nature needs some support and also some space. They roll up their sleeves, they donate their hard-earned resources and they raise the issues that count. Conservation organisations provide the foundation; a springboard of experience and expertise from which to operate: the Woodland Trust, the RSPB, The Wildlife Trusts, Butterfly Conservation, Froglife, Buglife, the British Trust for Ornithology, Plantlife, the National Trust and others are vital in protecting oaks and wider nature and in sharing information about them.

Young people
On my oak journey I met Sabaha Hussein, whose work on the Young People's Forest began when she was 17, along with fellow teenagers Elyse and Will. They have inspired rafts of young people and members of the local community to join them – every one of whom is falling a little bit more in love with nature – to help them and get more trees in the ground. As she embarked on a career in conservation later on, Elyse said, 'The Young People's Forest has contributed massively

towards the transformation of my confidence. I have gained the most valuable experience. I have had opportunities to learn, to physically take part and to meet new people.'

Campaigners

Back in time we saw Edward North Buxton inspire a movement that helped save Epping Forest and part of Hainault Forest. And we have to thank all the people who stood up and cried out when the government tried to sell off public forests in 2010, and who are still standing up for our trees and woods today.[24]

Now thank heavens for tree heroes up and down the country who are looking out for the threats to fight and the opportunities to champion. Some are adding vital voice and weight to let decision makers know how much they care about the environment and how we much we need our rules and regulations to reflect that. Others are working day in and day out in their locality to save the one, two or more trees that are valuable to them. In your own community you can see most clearly of all how every tree counts and every voice counts.

Charity staff and volunteers

Woodland Trust volunteer Kevin Stanley was so inspired by the Ancient Tree Inventory (ATI) that he set out on his own tree hunt, and has now found, measured and mapped more than 1,000 ancient and veteran trees – 300 of them oaks – that were previously unrecorded. Kevin is finding oak trees, hitherto unknown to conservationists, more than nine metres in girth.

Ruth Hyde, director of communications at the Woodland Trust says, 'There are thousands of conservation volunteers and staff across the UK, working hard day after day. They are getting on with it, often quietly, without expectation of great recognition, but each and every one of them deserves to be celebrated from the treetops because they are working to save our nature, our future.'

New royalty
Those age-old royal connections with the oak go on. King Charles III spearheaded Action Oak and the Queen's Green Canopy, calling on people in the UK to plant trees and networks of trees throughout the country to mark Queen Elizabeth II's 70 years on the throne. After planting an oak, he said, 'There is a reason for this profoundly symbolic act; planting a tree is a statement of hope and faith in the future.' The King is also highlighting the need to protect our ancients, announcing that 70 ancient trees – including The Curley Oak in Wentwood Forest – and 70 ancient woods are dedicated under the Queen's Green Canopy. He said, 'Forests are amongst the richest biological areas on earth. Tropical or temperate, they offer a magnificent variety of habitats for plants animals and microorganisms. Quite simply we cannot live without them.'[25]

Farmers and landowners
Here we can take inspiration from Charlie Burrell and Isabella Tree at Knepp; from James and Alka Hughes-Hallett planting in Somerset and inspiring people with not-so-galling gifts; or we can look to Lord and Lady Fellowes, who are not only protecting a magnificent old Turkey oak but are also planting oaks and other trees on their land for the future. Some own one acre, some own a thousand. These are the custodians of our trees, whether they be in fields, woodlands, hedgerows or towns.

Those gods of small things

Oh and the gods of small things, to borrow that phrase from Arundhati Roy again; all those scientists – formal and citizen – who are researching the facts, the fungi, the galls, the teeny, tiny wasps, the springtails, the lichens, the mosses. These are amazing people, who give the rest of us the evidence on which to base decisions. They sweat the small stuff so we don't have to and they bring out its beauty and its joy.

These are just a few of the groups we can name who are playing their part. There are so many more it is impossible to name them here, but let's champion them when we get the chance.

Conclusion

THE CREATION OF A THOUSAND
FOREST IS IN ONE ACORN.
RALPH WALDO EMERSON

IN MY OAK ADVENTURES I met people from all walks of life, up and down and east and west. They are all working to protect trees and the world they support. It is the love of all these people that will save the oak and our natural world for the future. I singled out some groups in Chapter 8, but truly there are too many to mention them all here – that's a good problem to have, of course.

So how about the rest of us? Are we oak heroes? If not, could we be? Yes, absolutely. The anthropologist Margaret Mead taught us not to doubt that 'a small group of thoughtful, committed citizens can change the world', pointing out that 'It's the only thing that ever has…'

We can work to protect the oak tree for the sheer love of this iconic plant. We can choose to save it for the wren that perches in it. We can opt to support the boletes that make an incredible risotto. We can choose to protect the oak and its world because it is tied up with the very health and happiness of our children. We can do it because there are 2,300 species – that we have counted so far – supported by the oak tree. We can do it because 300 of those species depend on UK oak trees for their very existence. We can do it because we know that it all links up – everything we do for nature impacts on something else and ultimately on us.

The next few pages contain suggestions on what you can do, how you can do it, and where you can do it.

Whatever you choose to do and however or wherever you choose to do it, for whatever reason, have fun – let's celebrate this amazing plant. Oh, and enjoy the cake.

Some Fun Stuff

HELPING LIFE BY ENJOYING LIFE
Share the love
My best tip is to enjoy the oak and its world. Visit some trees. Walk around a wood. Get to know a nearby tree if you can, watch it change through the seasons.

RSPB CEO Beccy Speight says, 'If you have learned to love a lichen or an oak or a long-tailed tit, then talk about it, with your family, with your friends. We need as many people as possible to understand and love oaks and their world because that will help us protect these treasures.'

And to know the oak and its world is to love it.

Become a citizen scientist
This isn't at all as difficult as it might sound and, if you have never done it, I recommend having a go. Mindful, absorbing, call it what you will. I have always found myself getting happily lost in the moment, other cares left behind.
- Watch a tree or three. Start recording within Nature's Calendar (naturescalendar.woodlandtrust.org.uk) and you'll be contributing to a biological record that dates back as far as 1736. The team give great guidance on getting started and the knowledge we have from the data gathered is invaluable.
- Join the ancient tree hunt. Get yourself a bit of oaky treasure. Map and measure ancient trees in your patch and add them to the Ancient Tree Inventory (ati.woodlandtrust.org.uk).
- See Observatree for information about becoming a specialist volunteer in spotting tree pests and diseases (observatree.org.uk).

- You need as little as 15 minutes to take part in the Big Butterfly Count each year (butterfly-conservation.org).
- Doing the annual Big Garden Birdwatch in January is an hour well spent – for both us and the birds (rspb.org.uk).

Join
The fantastic members and the people who donate regularly provide the absolute lifeblood for charities such as the Woodland Trust, The Wildlife Trusts, the RSPB, the National Trust, Plantlife, Buglife, Butterfly Conservation, the British Trust for Ornithology, and local wildlife and natural history groups like the Long Eaton Natural History Society (lensweb.wordpress.com) that led the moth hunt I went on.

Buy well
If you're buying trees, please buy only UKISG (UK/Ireland Sourced and Grown) to avoid importing tree disease. Consider a native tree, as they're usually the best option for wildlife. (See tips on growing your own oak below.)

Campaign
Signing up to a campaign newsletter can be a powerful act. Decision makers need to know we care deeply about nature. Every voice counts. Your MP cares about every vote. Save an ancient tree near you. Hold the people doing the 'blah blah blah' to account. I can tell you it feels really great when you win for nature and the future.

Volunteer
Check out the website of your nearest conservation organisation for details on how you can get involved. You don't have to be the type to tool up with saws and billhooks – although that can be great fun. While outdoorsy volunteers play their part, others can choose to support

conservation from a desk or an armchair by campaigning, donating and sharing key messages.

Use your superpower
Remember this superpower – consumer power – that we all have. Our choices make businesses listen, take notice, pick up the fight. Buying local and organic products helps nature. Dive in deeper, ask questions about the products you are buying, from your puppy food to your pension.

Do less and also do more
Drive less, fly less, cycle more, walk more. Let's take responsibility for every bit of carbon we emit. The things that reduce our carbon footprint help the natural world.

Tread lightly
The brilliant thing is that we can all enjoy nature, often for free. Many woods and nature reserves are free to visit. Go and love them and do, please, tread lightly; particularly around delicate treasures such as ancient trees – love them, enjoy them, stand back and enjoy the view of them. Splash your way through puddles on paths. Wearing wellies on muddy days in woods means you can avoid widening those paths and encroaching on sensitive spaces like bluebell patches.

Stop before you squash
This is an easy one. So often we see someone swipe an insect dead when it isn't even hurting them. Why take a life when you don't need to take a life? It might be a tiny little short life. But yes I'm talking about respect for the small stuff because everything, like everyone, has something amazing about it and has a vital role to play in the wider

world. That tiny beastie is in a food chain. It connects to the rest of the world, probably in multiple ways. It might even be like the gall wasps who played such a key role in human history. Finding the magic and beauty in tiny creatures can enrich our lives immeasurably. Thomas Hardy wrote about 'a longlegs, a moth, and a dumbledore' (bumblebee) as well as a 'sleepy fly that rubs its hands' in his poem 'An August Midnight': ' "God's humblest, they!" I muse. Yet why? They know Earth-secrets that know not I.'

Work with your furry friend
I love Pepe the pooch, likewise the beloved pet cats of my childhood, but I have to accept that furry friends can be destructive if left to their own devices. Keeping dogs on leads in sensitive areas makes a big difference to nature. Please follow guidelines and advice from conservation organisations. The RSPB has some great tips on helping keep garden wildlife safe from cats, for example planting prickly bushes beneath bird feeders. And remember, springtime and early summer are times at which lots of young birds and mammals are at their most vulnerable, still needing to build up their strength and survival skills.

Grow your own oak tree
Gardener, writer and TV presenter Danny Clarke gave me some advice on growing your own oak tree. Here are some of his top tips:
- The most fun way is to start off your oaks from acorns. Choose the nice firm ones, not the squishy jobs. Put them in water. Those that float are no good. Use those that sink.
- It is a good idea to sow your acorns in a pot first and cover them in mesh so your acorn doesn't get eaten by a squirrel.

- Location is key. Oaks grow in most soils, but you do need a big space for an oak as they can grow well over 30 metres (that could be five houses) tall. They're not ideal for small gardens (you do sometimes see them in smaller gardens but they need to be heavily cut back, a job you'd need to do regularly). So, unless you have lot of space, it may be a case of working with the leaders of a local community space, possibly a sports club or a school, who might be happy for you to plant an oak. Note also that the Woodland Trust, with the help of key supporters, offers free tree packs to schools and community groups.
- When you are planting out your tree, dig a hole bigger than the root ball and add some compost. Don't plant it too deep – that's a common mistake – you want the soil to cover the roots and not go up the trunk of the tree. If you have grown it in a pot you'll see where the soil should come up to. You might need to use a tree guard, especially if you are in an area with a lot of deer.
- The first three to five years of a tree's life are the most important. Don't let it dry out. Keep your tree well-watered.

Help an ailing tree
Arboriculturist Tony Kirkham of Kew Gardens fame gave me some advice on caring for a tree that isn't looking so well. Here are a few top tips I learned from Tony:
- If you have a tree and you're not sure what's wrong with it, put yourself in the tree's position and do a 360, look round and say, 'Well, if I was a tree what would be bothering me?', a bit like a vet. Because a tree can't say 'This branch is sore', or 'My roots aren't right'. This approach helped Tony – famously – work out how The Turner Oak's roots had become so compacted. It has also helped him diagnose

problems when trees have suffered from lawn weed-and-feed chemicals.
- Think about getting nature to work with you, instead of working against it. In a garden we often rake up leaves in order to have immaculate lawns and beds but think, instead, of the poor tree that has been there for 200 years, and which relies on dropping its leaves and recycling them itself. We can be quick to get rid of natural debris without putting anything back, and we wonder why trees are stressed. You can get mulching decks for mowers. They smash up the leaves, making life easier for the worms to take down. The worms will aerate the soil and de-compact the ground. It's amazing how fast worms work if you help them. If you must be tidy, perhaps compost your leaves and then put them back around the tree.
- Be careful when using strimmers around trees to avoid bruising the bark.

Go wild in a garden
The growth in interest in wildlife gardening is brilliant. And it all helps the world of the oak. Remember, for example, how the larvae of some beetle friends live in dead wood and yet their adults feed from flowers.

If you are lucky enough to have a green space to care for, don't forget Louise Hackett's log pile tip: top up those lovely logs every couple of years. Bury some of your logs in the ground – just leaving tree stumps where they are, half buried, helps. And Danny Clarke advised me during a chat, 'If a limb falls from a tree, leave it there if you can, cutting it up if you need to – it will help beetles and eventually return to the earth, enrich the soil and become part of the cycle.'
- My favourite gardening tip, and something we can do in a limited space, a windowsill perhaps, was given to

me by the gardener and TV presenter Chris Beardshaw: 'have a flower in flower every day of the year'.
- If you have space, think about planting for birds and other wildlife – roses, hollies, hawthorn – seed-and berry-bearing shrubs can make for wonderful wildlife meals, and they look beautiful.
- Think about the ways green patches link up. Can wildlife move to escape threats? Check to see if hedgehog highways have been blocked by a fenced-in garden, for example. Can you grow a hedge? Hedgerows – especially those using native plants – make wonderful wildlife corridors.
- Consider a pond. Just add water if you want it to buzz with wildlife. Even a buried washing up bowl will do, as long as creatures can climb in and out. Place a pile of stones at the edge to act as a stairway for fascinating visitors.
- Don't forget the power of sitting in a chair and just letting things go wild – marvellous. If you join in #NoMowMay – simply not mowing for a whole month – you'll be helping invertebrates. To paraphrase Plantlife, liberate your lawns, let the wildflowers bloom and provide a feast of nectar for our hard-working, hungry pollinators.[1]
- Finally, another lovely tip from Danny: 'A lot of people ask me to plant evergreen trees. I try to push back sometimes because to me, a tree not in leaf is even more beautiful than a tree with leaves. Think of those winter nights when the moon shines through the branches. It is good for the soul.'

HOW TO MEET A TREE

If you are used to going out into woods and beyond, this section is entirely unnecessary reading. However I'm conscious that getting into nature is by no means second nature for everyone. I was an adult when I first set out to visit a nature reserve and I was nervous. I wasn't sure what to expect, what to take with me. I worried about getting things wrong, causing offence in some way. Now I do it all the time, visiting nature is second nature. So here are a few tips for anyone who is setting out on a new nature journey.

Where to go and who to go with
- Hopefully there are trees near you so you can admire one in a park or on a (safe) roadside. Make a point of looking at it in different seasons; enjoying its wildlife – perhaps start with birds as they are often the easiest to see; or things that don't move, like moss and lichen.
- Be safe. If you are focusing in on nature please be aware of who else is about and what else is around you. Perhaps find a companion to join you in this activity.
- If venturing further afield, let someone know where you are heading to and when you expect to be back. Accidents are rare but you never know.
- Ideally start with group ventures. Look for events run by your local wildlife group, your Wildlife Trust, the Woodland Trust, RSPB, National Trust, etc.
- If you're not going as part of an organised group, you can usually find nearby woods and nature reserves by searching on the websites of conservation organisations. Before setting off, check out how easy they are to find and to access. This is crucial. Some are much easier than others with handy public transport or parking, but others are more remote, and harder to find, so do check out directions thoroughly.
- Give yourself plenty of time to ensure you are out of the wood or nature reserve before evening and darkness. While visiting a nature spot can be fantastic after dark – opening up more worlds – this isn't for beginners and definitely not something to do on your own. Perhaps hook up with

your local wildlife group or bat group for night-time nature adventures.
- Read through the Countryside Code: gov.uk/government/publications/the-countryside-code

Extras to check before a visit to a wood or nature reserve
- Directions and, if relevant, parking.
- The availability of mobile reception (some woods or nature reserves have none, so that is something to factor into your plan for the day).
- Whether there are sensitive areas and things to avoid trampling e.g. spring flowers such as bluebells and ancient or veteran trees.
- Which facilities are available, if any. They can range from a café, loos and some shelter from bad weather to often, well, just the wild stuff.
- The location of nearby cafés should you want refreshments, along with a loo stop after your visit if there are minimal facilities at the site.
- Whether, if relevant to you, there is a policy about dogs (often they are welcome on leads, but in some places they aren't invited for good reasons).
- Check out the weather forecast and ensure you are prepared for sun or rain – ideally both. Hats are top of the equipment list to protect you either from sunshine or the cold, or even both on the same day, and protection from the sun is of course vital on hot days.

Must-takes
- Good footwear. Walking boots are ideal. Wellies are important in muddy conditions in sensitive areas so you don't end up avoiding the paths and trampling delicate flowers and other wildlife.

- Outdoor clothing for protection from sun and rain. Hat, gloves and a change of socks are good to have on hand. Umbrella if you are an umbrella kind of person – I know I am.
- Sun protection in hot weather.
- Refreshments – drinking water is a must. Also consider snacks or a picnic and a flask with a hot drink. Maybe a slice of cake.
- Bag for litter.
- Dog walkers: lead, poo bags and dog drinking water.
- Hand sanitiser.
- Map, especially if visiting a large reserve or area of the countryside.
- Fully charged mobile phone.
- First-aid kit – for simplicity you can get a small version that you can just hang off a belt.

Optional extras
- A book – or, often even better, a light, fold-out guide – about wildlife you are especially interested in. I love the fold-out wildlife guides from the Field Studies Council (FSC) (field-studies-council.org)
- Binoculars
- Hand lenses can be great for looking more closely at twigs, buds, galls, lichens, etc. You can get a cheap, basic one to fit in a wallet. And don't forget you can usually magnify little things using a smartphone.
- A picnic blanket. Or, alternatively, a sitting mat. I was inspired by the removable seat pad inside my little backpack to make an extra version for a friend out of some old packaging material (the bendy, foamy type – you can just cut it to fit nicely in your bag). These are handy for avoiding a damp bottom when sitting on a soggy log.

A TREE FOR ALL SEASONS

The best season is the one we are in. Yes, even winter. While it is easy to spend dark days yearning for spring and summer there is always something to get excited about in the life of the oak and also, of course, the wider tree world. There is always something to be mindful about, always something to get outside for. Here are some ideas to get you started, no matter the month.

Winter

This is a time for tuning in and seeing things you don't usually see.

- A great time to start learning birdsong. There isn't so much going on and you can tune in your ears to the birds that are singing throughout the year. Watch a robin. Tune in to its song. Follow Julian Branscombe's tips in Chapter 1 and build up your repertoire as the winter turns into spring.
- Winter is lichens' and bryophytes' time to shine. Enjoy their shapes and textures. If you'd like to go further, arm yourself with a Field Studies Council guide or check out some images from the Internet and start getting to grips with oak moss and its friends.
- This is the time we can enjoy the trees in their leafless glory; see the twists and turns of branches and imagine how the ship builders would have worked with their shapes.
- It is a fun time for seeing tree faces with kids and big kids – when the arrangement of grooves and notches in the bark looks like a person or a creature staring back at you. I have even seen a site on social media dedicated to pictures of these. And if you can't find a tree face, you can always make your own: Helen Hampton, education ranger at Essex Country Parks recommends this as an activity with

children. Helen says, 'Slap some mud onto the tree trunk and press leaves into it to make the iconic green man – or woman – face.'
- Try Brian Eversham's great tip: in late winter tie a ribbon on an oak twig, somewhere you visit it as the seasons change. Revisit it, photograph it if you can. This is a lovely lens through which to see a tree. It is surprising how quickly things change once spring gets going.
- Gardeners: if you have a bit of outside space, consider your birds. We have seen how seed-bearing plants can help. And if you go down the route of putting out bird feeders, it is vital to clean feeders frequently to prevent disease; check out more advice from wildlife organisations. If you can have a flower in flower it will help insects that emerge on warm days.

Spring

Wildlife flirting time. It's all getting a bit sexy now.
- Enjoy the drumming of woodpeckers and the chiff-chaffing of, well, chiffchaffs, while you can before the full force of the bird orchestra builds up. Try to hold onto its coat-tails and join in the International Dawn Chorus Day.[2]
- Birdcams and other wildlife cams can make great spring viewing from your armchair.
- Oak budburst tends to start around late March and can go on until early May. It will depend on how far north or south you are. It is followed by first leaf and first flowering. Tune in to the Woodland Trust's Nature's Calendar for more information.[3]
- Time for woodland flowers and wellies. A bluebell wood in April or May will do you good, I promise. If I could prescribe this on the NHS I would. There is nothing like the beauty of

a bluebell wood when it looks like a sea of blue, and that smell! It is amazing. Bluebell health warning: please, please stick to the paths to avoid trampling these precious flowers. Don't forget your wellies.
- Late spring is a good time to start beetling about. With the blooming of flowers come amazing emerging insects, which in turn provide feasting for baby birds. Check out those longhorn beetle friends.

Summer

Well, it's all gone wild now.
- This is the perfect season for 'five-to-nine-ing'. Think not 'nine to five' but 'five p.m. to nine a.m.' because we have light, glorious light. Get out in the long summer evening and return to your day life refreshed and full of nature. A great time to refocus.
- Galls galls galls. Early summer. This is a great moment for oak apples and other galls. See Chapter 2 for ways to get galling.
- '30 Days Wild' is The Wildlife Trusts' annual challenge, where they ask everyone to do one wild thing a day throughout the month of June.[4]
- It is butterfly time – join a nature walk to spy a purple hairstreak. Join in the Big Butterfly Count.[5] Be inspired by Butterfly Conservation and Buglife.
- Keep an eye on your cats and dogs to avoid fledgling casualties.
- Midsummer to autumn – watch the baby acorns grow and fatten up on the oaks.
- At dusk, enjoy the silhouettes of bats around hedges and woodland edges.
- Midsummer nights' adventures – take a late-evening walk in a wood with friends. Be safe.

Autumn

Keats's 'Season of mists and mellow fruitfulness…'

- Fungal forays time. See Chapter 6 for some oak-related fungal inspiration.
- Think about the world beneath your feet – all that wonderful underground networking
- In mast years – those when the acorns are abundant – collect an acorn and watch it grow in an acorn vase (they have a bulbous base and a narrow neck with a squeeze in it to stop the acorn falling down) just to see how it develops.
- This is a good time to spot jays in woodlands, as they are working hard collecting their acorns. You might also see mammals like squirrels bounding about, creating stashes of nuts for the oncoming winter.
- There is plenty of bird fun to be had. There is movement and migration going on. Pied flycatchers, for example, set out on their big journeys around September. Meanwhile tawny owls get wonderfully noisy during October and November and are particularly active *twit-twoo*-ing as they call out their territory in woods and parks.
- Increasingly naked oaks reveal birds' nests and squirrel dreys and their own beautiful limbs.
- Enjoy the humble tree climber, ivy. This is its time to shine as its late flowers provide late nectar for insects that are still hanging about. On a warm day stand under an ivy flower-filled oak tree and listen to the sounds of the insects.

Whatever you do – or, in the case of wildlife gardeners perhaps, don't do – I hope you enjoy the oak and its world because there is so much delight and wonder to be had right there.

Wishing you some happy oakventures, Jules

Acknowledgements

ONE OF THE GREAT THINGS about hanging out in the worlds of woods and wildlife is that people are generous with their knowledge. They are also, invariably, kind, fascinating and fun. I can't begin to thank everyone enough, including the following special people, for their support, their expertise, their time and their brainpower. And I'm now really worried about missing someone out. I've probably left someone out... I've almost certainly missed someone out. If that is you I am sorry – please get in touch and berate me and I hope I can make amends somehow, perhaps with some cake. And thank you.

Thank you to Greystone's lovely Rob Sanders for inviting me to write this book and to all the wonderful Greystone team for your support, patience, encouragement and great wisdom: Fiona Brownlee, Lesley Cameron, Jennifer Croll, Nancy Flight, Andrew Furlow, Elizabeth Peters and Jessica Sullivan.

Endless thanks to my talented and always-supportive sister, Sally Mollan, for the beautiful illustrations, and to those who kindly allowed her to use their photographs for inspiration: Christoph Benisch, Graham Calow and Andy Lindsley.

Thank you to all those who have shared their guidance, time, knowledge, expertise and inspiration: Nick Atkinson, Tamsin Betti, Lynne Boddy, Richard Blanchard, Toby Blower, Julian Branscombe, Marion Bryce and members of LENS Wildlife Group, Abi Bunker, Danny Clarke, Carl Cornish,

Alan Crawford, Irena Ekart, Brian Eversham, Emma Fellowes, Alasdair Firth, Emma Gilmartin, Rebecca Gosling, Louise Hackett, Helen Hampton, Lyn Hartman, James Heal, Russ Hedley, Alex Hollington, LooLoo Holloway, Alka Hughes-Hallett, James Hughes-Hallett, Sabaha Hussein, Ruth Hyde, James Jesson, Jonathan Jukes, Tony Kirkham, Nick Littlewood, Karl Mitchell, Ruth Mitchell, Darren Moorcroft, Jim Mulholland, Eben Neale, Chris Nichols, Ken Orpe, Pat Orpe, Tom Reed, Chloe Ryder, Jim Smith-Wright, Jane Southey, Beccy Speight, Kevin Stanley, Andy Taylor, Fiona Theokritoff, Elyse White, Lorienne Whittle, Sarah Willis and John Wright

Thank you to family and friends for bearing with me when I realised writing a book was a lot harder than I expected, especially to the incredible Toby Blower for keeping me company on many oaky walks along with Pepe the pooch; for helping me find the fun in the world and for being there every day in every way.

Finally, a big shout out to our conservation organisations, to those who work for them and volunteer for them and to every single person who has ever supported them because you, you nature-saving friends, are doing something both vital and brilliant.

Endnotes

INTRODUCTION

1 Dr Ruth Mitchell researched life supported by oak trees along with a team of fellow experts in a project for PuRpOsE (PRotect Oak Ecosystems). R. J. Mitchell, P. E. Bellamy, C. J.Ellis, R. L. Hewison, N. G. Hodgetts, G. R. Iason, N. A. Littlewood, S. Newey, J. A. Stockan, A. F. S. Taylor, 'Collapsing foundations: the ecology of the British oak, implications of its decline and mitigation options', *Biological Conservation*, 233 (May 2019), 316–327. DOI 10.1016/j. biocon.2019.03.040.

   See also Action Oak, 'PuRpOsE: uncovering the biodiversity of oak trees', actionoak.org/projects/purpose-uncovering-biodiversity-oak-trees

2 Brian Eversham is chief executive at the Wildlife Trust for Beds, Cambs & Northants, wildlifebcn.org/our-people-and-partners/our-staff

3 Action Oak, 'About', actionoak.org/about

4 Action Oak, 'About', actionoak.org/about

5 Professor Richard Ellis and Dr Jo Clark, 'Enhanced acorn production to regenerate native oak woodlands in the UK', 2023, actionoak.org/projects/enhanced-acorn-production-regenerate-native-oak-woodlands-uk

6 See the Woodland Trust's handy guide, 'A–Z of British Trees' at woodlandtrust.org.uk/trees-woods-and-wildlife/british-trees/a-z-of-british-trees/

7 Encyclopaedia Britannica, 'Oak', 20 October 2023, britannica.com/plant/oak

   See also Rachelle Dragani, 'How Many Types of Oak Trees Are There?', 22 November 2019, sciencing.com/many-types-oak-trees-there-5347784.html

   See also Georgette Kilgore, 'Oak Tree Leaf Identification Chart With Locations (13 Oak Species)', 4 May 2023, 8billiontrees.com/trees/oak-tree-leaf-identification-chart/

8 Woodland Trust, 'Oak', ati.woodlandtrust.org.uk/how-to-record/species-guides/oak/
9 Aljos Farjon, *Ancient Oaks in the English Landscape* (Kew Publishing, London, 2017), 184
10 Woodland Trust, 'WE CREATE', woodlandtrust.org.uk/about-us/what-we-do/we-plant-trees/
11 Simon King, 'Tree ID – Telling an Oak from an Ash', 2013, youtube/373vayTmJ-w
12 The Wildlife Trusts, 'Sessile oak', wildlifetrusts.org/wildlife-explorer/trees-and-shrubs/sessile-oak
13 Woodland Trust, 'State of the UK's Woods and Trees 2021', woodlandtrust.org.uk/media/51705/state-of-the-uks-woods-and-trees-2021-thewoodlandtrust.pdf page 45
    See also Woodland Trust, 'Ancient Tree Inventory', ati.woodlandtrust.org.uk/
    See also Action Oak, 'Did you know? Here are some amazing facts and figures about the Great British oak!', actionoak.org/did-you-know
14 Ancient Tree Forum, ancienttreeforum.org.uk/

1: DOORWAYS TO OTHER WORLDS

1 For those less acquainted with the music of the 1980s, 1990s and beyond, Phil Collins is a drummer and singer-songwriter, perhaps most famously in the band Genesis. Phil Collins (@officialphilcollins) instagram.com/officialphilcollins/
2 Gerard Gorman, *Woodpeckers* (London: RSPB Spotlight; Bloomsbury Wildlife, 2018), 5
3 Gorman, *Woodpeckers*, 35
4 Jean Parrott, 'The Great Spotted Woodpecker', British Trust for Ornithology, 21 July 2014, bto.org/sites/default/files/u23/images/Ambassadors/jean_parrott_articles/The%20Great%20Spotted%20woodpecker%20%281%29.pdf
5 Parrott, 'The Great Spotted Woodpecker'
6 Emily Anthes, 'Woodpecker inspires cardboard bike helmet', BBC Future, 18 November 2014, bbc.com/future/article/20130115-woodpecker-inspires-bike-helmet
7 'The woodland drummers', Cornwall Wildlife Trust blog, 1 February 2023, cornwallwildlifetrust.org.uk/blog/thewildlifetrusts/woodland-drummers
8 The Woodpecker Network, woodpecker-network.org.uk/
9 Amy Lewis, 'British Woodpeckers: Identification Guide and Calls',

Woodland Trust blog, 30 September 2021, woodlandtrust.org.uk/blog/2021/09/british-woodpecker-id/
10 Gorman, *Woodpeckers*, 48–49
11 Birdnote, 'This Bird Will Put a Spell on You', Audubon, 1 March 2014, audubon.org/news/this-bird-will-put-spell-you
12 British Trust for Ornithology, BirdFacts, 'Wryneck', bto.org/understanding-birds/birdfacts/wryneck
13 Gorman, *Woodpeckers*, 5
14 Gorman, *Woodpeckers*, 31
15 Gorman, *Woodpeckers*, 33
16 Stump Up For Trees, stumpupfortrees.org/
17 British Trust for Ornithology, BirdFacts, 'Pied Flycatcher', bto.org/understanding-birds/birdfacts/pied-flycatcher
18 Bird Guide, British Garden Birds, '(Eurasian) Nuthatch', garden-birds.co.uk/birds/nuthatch.html
   See also Amy Lewis, 'Bird Song Identification: Songs And Calls For Beginners, Nuthatch (Sitta europaea)', Woodland Trust blog, 10 April 2019, woodlandtrust.org.uk/blog/2019/04/identify-bird-song/
19 M. G. Morris and F. H. Perring, eds, *The British Oak* (published for the Botanical Society of the British Isles by E. W. Classey Ltd, 1974) 182–184
20 John Lewis-Stempel, *The Glorious Life of the Oak* (New York: Doubleday, 2018), 49
21 Woodland Trust, 'Lesser Spotted Woodpecker (*Dryobates minor*)', woodlandtrust.org.uk/trees-woods-and-wildlife/animals/birds/lesser-spotted-woodpecker/
22 *Get Birding*, Series 1: Episode 4 – Samuel West, Lucy Lapwing, Sam Lee, 12 March 2021, https://shows.acast.com/get-birding/episodes/episode-4
23 Dominic Couzens, www.birdwords.co.uk
24 Gorman, *Woodpeckers*, 88
25 Woodland Trust, 'The Major Oak', ati.woodlandtrust.org.uk/
26 Walking Englishman, 'Edwinstowe and Sherwood Forest', walkingenglishman.com/outandabout/centralengland/03edwinstowe.html
27 BBC, 'Suffragette Oak in Glasgow is named Scotland's Tree of the Year', 28 October 2015, bbc.co.uk/news/uk-scotland-glasgow-west-34655433
28 Ellen Castelow, 'The English Oak', historic-uk.com/CultureUK/The-English-Oak/
29 Archie Miles, *The British Oak* (London: Constable, 2016), 175
30 Miles, *The British Oak*, 99

31 BBC Wales History, 'Owain Glyndwr', August 2009, bbc.co.uk/wales/history/sites/themes/figures/owain_glyndwr.shtml
32 Miles, *The British Oak*, 178
33 Miles, *The British Oak*, 94
34 Visit Sherwood Forest, 'National Tree Week Man Eating Caterpillar Day 5', 28 November 2018, www.facebook.com/watch/?v=5254 11191311943
35 Miles, *The British Oak*, 8–9

2: WASPS, WORDS AND OAKSPIRATIONS

1 Alison Archibald, 'Galling ink! How The National Archives preserves millions of documents written in iron gall ink', The National Archives blog, 7 March 2021, blog.nationalarchives.gov.uk/galling-ink-how-the-national-archives-preserves-millions-of-documents-written-in-iron-gall-ink/

See also George McGavin, '2/2 The Oak Tree, Natures Greatest Survivor – March to August', youtube.com/watch?v=hQNMceu9ZmM

2 Paul Garside and Zoë Miller, 'Iron gall ink on paper: Saving the words that eat themselves', British Library blog, 3 June 2021, blogs.bl.uk/collectioncare/2021/06/iron-gall-ink-on-paper-saving-the-words-that-eat-themselves.html

3 Ben King, 'Life cycle of the marble gall wasp Andricus kollari', youtube.com/watch?v=8hjrkUo3wVE

4 Woodland Trust, 'Oak, Turkey (Quercus cerris)', woodlandtrust.org.uk/trees-woods-and-wildlife/british-trees/a-z-of-british-trees/turkey-oak/

5 Woodland Trust, 'Oak marble gall wasp, Scientific name: Andricus kollari', wildlifetrusts.org/wildlife-explorer/galls/oak-marble-gall-wasp

6 Amateur Entomologists' Society, 'True bugs (Order: Hemiptera)', amentsoc.org/insects/fact-files/orders/hemiptera.html

7 Oxford Reference, 'Royal Oak Day, Oak Apple Day', oxfordreference.com/display/10.1093/oi/authority.20110803100431700

8 Dr Willem N. Ellis, 'Leafminers and plant galls of Europe', bladmineerders.nl/

9 Institute of Chartered Foresters, 'New Tree Charter to Launch at Lincoln Castle', 5 September 2017, charteredforesters.org/new-tree-charter-launch-lincoln-castle

10 Bowthorpe Park Farm, 'The Tree', bowthorpeparkfarm.co.uk/the-tree/

11 Simon Ingram, 'Striking Artworks Capture the Majesty and Magic of the Ancient Oak Tree', 22 February 2019, nationalgeographic.co.uk/environment/2019/02/striking-artworks-capture-the-majesty-and-magic-of-the-ancient-oak-tree
12 Encyclopaedia Britannica, 'Herne The Hunter', britannica.com/topic/Herne-the-Hunter
13 VisitScotland, 'The Birnam Oak', visitscotland.com/info/see-do/the-birnam-oak-p2571371
14 Scott Smith, 'The Theology of Ents (Lord of the Rings)', 30 December 2009, thescottsmithblog.com/2009/12/man-eating-forests-tolkien-shakespeare.html
15 Undiscovered Scotland, 'Birnam Oak', undiscoveredscotland.co.uk/dunkeld/birnamoak/index.html?utm_content=cmp-true
16 Miles, *The British Oak*, 129
17 Kylie Harrison Mellor, 'Ancient Yew Trees: The UK's Oldest Yews', Woodland Trust blog, 22 January 2018, woodlandtrust.org.uk/blog/2018/01/ancient-yew-trees/
18 Romantic Circles, 'William Cowper, "Yardley Oak" ', romantic-circles.org/editions/poets/texts/yardleyoak.html
19 Joe Hamer, 'Inside DIO, Yardley Chase Training Area: Northamptonshire's hidden gem', Defence Infrastructure Organisation blog, 27 February 2023, insidedio.blog.gov.uk/2023/02/27/yardley-chase-training-area-northamptonshires-hidden-gem/
20 James Canton, *The Oak Papers* (Canongate Books Ltd, Edinburgh, 2020), 143

3: CROWNING GLORIES

1 National Trust, 'Kedleston Hall', nationaltrust.org.uk/visit/peak-district-derbyshire/kedleston-hall
2 Butterfly Conservation, 'Life cycles of Purple Hairstreak and White-letter Hairstreak', butterfly-conservation.org/sites/default/files/2020-08/Purple%20and%20white-letter%20Hairstreak.pdf
    See also Butterfly Conservation Yorkshire, 'Purple Hairstreak *Neozephyrus quercus* (Linnaeus 1758)', yorkshirebutterflies.org.uk/yorkshire-species/purple-hairstreak
3 Woodland Trust, 'Young People's Forest at Mead', woodlandtrust.org.uk/visiting-woods/woods/young-peoples-forest-at-mead/
4 #iwill, iwill.org.uk/
5 Butterfly Conservation, 'The State of the UK's Butterflies 2022', butterfly-conservation.org/sites/default/files/2023-01/State%20of%20UK%20butterflies%202022%20report.pdf page 8

6 Butterfly Conservation, 'Occasionally troublesome moths', butterfly-conservation.org/moths/why-moths-matter/occasionally-troublesome-moths
7 mothnight.info/about-moth-night/
8 Butterfly Conservation, 'Peppered Moth and natural selection', butterfly-conservation.org/moths/why-moths-matter/amazing-moths/peppered-moth-and-natural-selection
9 Arizona State University, 'Peppered Moths: Natural Selection in Action', askabiologist.asu.edu/peppered-moths-game/play.html
10 *More or Less*, 'Delta cases, blue tits and that one-in-two cancer claim', (BBC Radio 4, 23 June 2021) bbc.co.uk/programmes/m000x4vn
11 RSPB, 'Helping birds near you', rspb.org.uk/birds-and-wildlife/advice/how-you-can-help-birds/feeding-birds/
12 Richard Jones, 'Honeydew: what it is and why ants love it so much', 20 October, 2022, discoverwildlife.com/animal-facts/insects-invertebrates/honeydew/
13 Patrick Barkham, 'Ants run secret farms on English oak trees, photographer discovers', *Guardian*, 24 January 2020, theguardian.com/environment/2020/jan/24/ants-run-secret-farms-on-english-oak-trees-photographer-discovers
14 Big Butterfly Count, bigbutterflycount.butterfly-conservation.org/
15 mothnight.info/about-moth-night/
16 John Lewis-Stempel, *The Glorious Life of the Oak* (Doubleday Publishers, London, 2018), 15
17 Lewis-Stempel, *The Glorious Life of the Oak*, 26
18 Julian Hight, *Britain's Tree Story* (National Trust Books, 2011), 35
19 Lewis-Stempel, *The Glorious Life of the Oak*, 21
20 National Portrait Gallery, 'William Carlos (Careless) (died 1689), Royalist soldier', npg.org.uk/collections/search/person/mp00758/william-carlos-careless
21 Authorship unclear, possibly written by John Wade around 1660.
22 Eloise Feilden, 'Top 10 most popular pub names in the UK', 1 June 2022, thedrinksbusiness.com/2022/06/top-10-most-popular-pub-names-in-the-uk/
23 English Heritage, 'The Royal Oak', english-heritage.org.uk/visit/places/boscobel-house-and-the-royal-oak/things-to-do
24 The Diary of Samuel Pepys, Friday 1 June 1660, pepysdiary.com/diary/1660/06/01/
25 'Oak Apple Day', nationaltoday.com/oak-apple-day
26 Grosvenor Prints Catalogue, ilab.org/assets/catalogues/Grosvenor-Prints-Catatalogue-124.pdf page 9

27 Miles, *The British Oak*, 291
28 Hight, *Britain's Tree Story*, 24
29 Hight, *Britain's Tree Story*, 54–55
30 As quoted in Miles, *The British Oak*, 47
31 Oliver Rackham, *Woodlands* (Glasgow: William Collins, 2015), 129

4: LIFE, DEATH AND BEETLING ABOUT

1 Rackham, *Woodlands*, 128
2 Liam Olds, 'Back from the Brink: In pursuit of rare saproxylic invertebrates', Buglife, naturebftb.co.uk/wp-content/uploads/2021/09/In-pursuit-of-rare-saproxylic-invertebrates-Liam-Olds.pdf
3 The Wildlife Trusts, 'Black-and-yellow longhorn beetle', wildlifetrusts.org/wildlife-explorer/invertebrates/beetles/black-and-yellow-longhorn-beetle
4 UK Beetle Recording, 'About Longhorn Beetles', coleoptera.org.uk/cerambycidae/about-longhorn-beetles
5 Helen Keating, 'Giant Hogweed: The Facts', Woodland Trust blog, 29 Jun 2022, woodlandtrust.org.uk/blog/2022/06/giant-hogweed-facts/
6 See field-studies-council.org/shop/publications/longhorn-beetles-guide/
  And there's a ton of information on coleoptera.org.uk/cerambycidae/home
7 UK Beetles, 'Dorcus parallelipipedus (Linnaeus, 1758) Lesser Stag Beetle', ukbeetles.co.uk/dorcus-parallalopipedus
  See also The Wildlife Trusts, 'Scientific name: Dorcus parallelipipedus', wildlifetrusts.org/wildlife-explorer/invertebrates/beetles/lesser-stag-beetle
8 Woodland Trust, 'State of the UK's Woods and Trees 2021', page 13
9 Buglife, 'Welsh Blue Ground Beetle Project', buglife.org.uk/projects/blue-ground-beetle-2/
10 BBC, 'Sherwood Forest photographic exhibition to celebrate anniversary', 18 October 2014, bbc.co.uk/news/uk-england-nottinghamshire-29672962
11 Woodland Trust, 'State of the UK's Woods and Trees 2021', page 227
12 W. H. Auden, *Bucolics*, 'II. Woods'
13 Eleanor Clark, 'What's the difference between a wood and a forest?', Woodland Trust blog, 21 Mar 2018, woodlandtrust.org.uk/blog/2018/03/difference-between-wood-and-forest/

14 Isabella Tree, *Wilding: The Return of Nature to a British Farm* (Picador, London, 2018), 124
15 Rackham, *Woodlands*, 107
16 The Wildlife Trusts, 'Living Landscapes', scottishwildlifetrust.co.uk/our-work/our-projects/living-landscapes/ and wildlifetrusts.org/sites/default/files/2018-11/A%20living%20landscape%20%28full%20report%29.pdf
17 Victoria Nolan, Francis Gilbert, Tom Reed, Tom Reader, 'Distribution models calibrated with independent field data predict two million ancient and veteran trees in England', *Ecological Applications*, 32:8 (22 June 2022), doi.org/10.1002/eap.2695
18 Woodland Trust, 'Ancient Tree Inventory'
19 Woodland Trust, 'State of the UK's Woods and Trees 2021', page 48
20 Woodland Trust, 'State of the UK's Woods and Trees 2021', page 39
21 Woodland Trust, *Wood Wise*, December 2019, woodlandtrust.org.uk/publications/2019/12/wood-wise-life-in-deadwood/ pages 12–14
22 Forest Research, 'Oak Pinhole Borer (Platypus cylindrus)', forestresearch.gov.uk/tools-and-resources/fthr/pest-and-disease-resources/oak-pinhole-borer-platypus-cylindrus/

5: ENCHANTED FORESTS

1 Roman Britain, 'The Druids and Druidism', roman-britain.co.uk/the-celts-and-celtic-life/the-druids-and-druidism/
2 Sir James George Frazer, *The Golden Bough: A Study in Magic and Religion* (Mineola, NY: Dover Books, 2003; first published in 1906 and 1915, abridged in 1922), 128
3 Frazer, *The Golden Bough*, 128
4 Frazer, *The Golden Bough*, 129
5 Icy Sedgwick, 'Meet the Oak, the Favoured Tree of the Forest!', 4 September 2021, icysedgwick.com/oak-tree-folklore/
6 Frazer, *The Golden Bough*, 94
7 Frazer, *The Golden Bough*, 129
8 *You're Dead To Me* 'Medieval Christmas', (BBC Radio 4, 25 December 2021) bbc.co.uk/programmes/p0b97gdq
9 Exploring London, 'What's in a name?... Gospel Oak', September 16 2019, exploring-london.com/tag/john-wesley/
10 Miles, *The British Oak*, 83–84
11 Robin Harford, *Oak Notebook* (Amazon, 2018), 43
12 Margaret Baker, *Discovering the Folklore of Plants*. (Shire Publications, Bloomsbury Publishing plc, Oxford, 2008), 112

13 Icy Sedgwick, 'Oak trees and the Weather', icysedgwick.com/oak-tree-folklore/
14 Harford, *Oak Notebook*, 37
15 'Saving the Balfron Oak', balfron.org.uk/tag/clachan/
16 Ron Porley and Nick Hodgetts, *Mosses and Liverworts*, The New Naturalist Library (Collins, London 2005), 1
17 Michael Marshall, 'First land plants plunged Earth into ice age', 1 February 2012, newscientist.com/article/dn21417-first-land-plants-plunged-earth-into-ice-age/
18 Royal Botanic Garden Edinburgh, 'Bryophytes', rbge.org.uk/science-and-conservation/herbarium/our-collections/bryophytes/
19 Mae Clair, 'Mythical Monday: The Moss People by Mae Clair', 29 September 2014, maeclair.net/2014/09/29/mythical-monday-the-moss-people-by-mae-clair/
20 Lisa Schneidau, *Botanical Folk Tales of Britain and Ireland* (History Press, Cheltenham, 2018), 28–29.
21 Public Library of Science, 'Alongside Otzi the Iceman: a bounty of ancient mosses and liverworts', 30 October 2019, phys.org/news/2019-10-otzi-iceman-bounty-ancient-mosses.html
22 Sean R. Edwards, 'English Names for British Bryophytes', British Bryological Society Special Volume No. 5, 2020, britishbryologicalsociety.org.uk/wp-content/uploads/2021/01/English_Names-5.01-Edwards-Sean-2020.pdf
23 Porley & Hodgetts, *Mosses and Liverworts*, xi
24 John Craven, *John Craven's Countryfile Handbook* (BBC Books, London, 2010) 205
25 Rebecca Yahr, 'Lichen identification Workshop for beginners', Royal Botanic Garden Edinburgh with the Botanical Society of Britain and Ireland, November 19 2020, youtube.com/watch?v=qJa1olj3vRs
26 Oliver Gilbert, *Lichens* (London: HarperCollins, 2000), 28
27 Gilbert, *Lichens*, 28
28 Frank S. Dobson, *Lichens: An Illustrated Guide to the British and Irish Species* (published by the British Lichen Society in collaboration with the Richmond Publishing Co. Ltd, 2018), 6
29 Dobson, *Lichens: An Illustrated Guide*, 6
30 Gilbert, *Lichens*, 23
31 UK AIR Air Information Resource, 'Glossary', uk-air.defra.gov.uk/air-pollution/glossary.php?glossary_id=2
32 British Ecological Society, 'We Need to Talk About Nitrogen', britishecologicalsociety.org/need-talk-nitrogen/

33 Gilbert, *Lichens*, 69
34 Encyclopaedia Britannica, 'Oak moss', britannica.com/science/oak-moss
35 Gilbert, *Lichens*, 50
36 Mark Powell @obfuscans3, February 2022, twitter.com/obfuscans3/status/1488549005311332353?lang=en-GB

6: INCREDIBLE EDIBLES

1 First Nature, 'Fistulina hepatica (Schaeff.) With. - Beefsteak Fungus', first-nature.com/fungi/fistulina-hepatica.php
2 For example the National Trust's 'Foraging for wild food', nationaltrust.org.uk/who-we-are/about-us/our-policy-on-foraging-for-wild-food
3 Team Candiru, 'Acorn Weevils do the Oaky Pokey', youtube.com/watch?v=2ftGzcuNawM
4 Extract from Hesiod's *Works and Days*, from William Bryant Logan, *Oak: The Frame of Civilization* (New York: Norton, 2006), 35
5 'Bread From Acorns (1933)', the Orgone Archive, youtube.com/watch?v=x4-F5N63Cdo. Ta-bu-ce was also known as Maggie Howard.
6 acorncoffeeflour.com Note, if ordering from the UK you will need to check any goods sent outside the UK are declared: gov.uk/goods-sent-from-abroad
7 Taste of the Wild, 'Quercus alba, White Oak', bio.brandeis.edu/fieldbio/Edible_Plants_Ramer_Silver_Weizmann/Pages/spp_page_oak.html
8 Logan, *Oak: The Frame of Civilization*, 48
9 William Cobbett, *Rural Rides*, gutenberg.org/files/34238/34238-h/34238-h.htm, 30
10 Harford, *Oak Notebook*, 23
11 Richard Mabey, *Plants with a Purpose* (Glasgow: Collins, 1977), 82
12 John Wyatt, *The Shining Levels* (Random House, London, 1993, first published 1973), 6.
13 Neil Prior, 'David Lloyd George's WWI act "helped save whiskey"', 13 March 2016, bbc.co.uk/news/uk-wales-35790275
14 'The Barrel Effect: Why Oak casks have stood the test of time', *The Food Programme* (BBC, 14 March 2021), bbc.co.uk/programmes/m000t409
15 Jennifer Griffin, 'The Life of a Bourbon Barrel', 17 May 2018, speysidecoopergaeky.com/bourbon-barrels/1021/

16 Terry Yarrow, 'The Remedy Oak, a Natural Cure!', 11 June 2017, thedorsetrambler.com/2017/06/11/the-remedy-oak-a-natural-cure/
17 Harford, *Oak Notebook*, 15
18 Fiann Ó'Nualláin, *The Holistic Gardener: Natural Cures for Common Ailments* (Cork: Mercier Press, 2016), 192
19 Harford, *Oak Notebook*, 21–23
20 Woodland Trust, 'Chicken of the Woods (Laetiporus sulphureus)', woodlandtrust.org.uk/trees-woods-and-wildlife/fungi-and-lichens/chicken-of-the-woods/
21 Jamie Kunka, 'Chicken Of The Woods Mushroom – Identification, Facts and Tasty Sandwich', youtube.com/watch?v=iMeKyuxFLaQ
22 The Forest Pharmacy, 'Chicken-of-the-Woods: Medicinal Benefits', theforestfarmacy.com/chicken-of-the-woods-mushroom-medicinal-benefits
23 Tree Council, 'Why trees are good for you', 2020, treecouncil.org.uk/wp-content/uploads/2020/02/Why-trees-are-good-for-you.pdf
24 'Resilient Young Minds, a green social prescribing pilot', 13 July 2021, facebook.com/finglewoods/videos/677998423168727
25 Woodland Trust, 'The New Forest with Adam Shaw and Dan Snow', *Woodland Walks* podcast, 12 March 2020, uk-podcasts.co.uk/podcast/woodland-walks-the-woodland-trust-podcast/13-the-new-forest-with-dan-snow
26 Mark Stanford, 'Dr Amir Khan on podcast about nature and our mental health', *Bradford Telegraph and Argus*, 26 February 2021, thetelegraphandargus.co.uk/news/19118263.dr-amir-khan-podcast-nature-mental-health/
27 Action Oak, 'PuRpOsE: uncovering the biodiversity of oak trees', actionoak.org/projects/purpose-uncovering-biodiversity-oak-trees
28 'Oak Polypore', wildfooduk.com/mushroom-guide/oak-polypore/
29 Woodland Trust, 'Beefsteak Fungus', woodlandtrust.org.uk/trees-woods-and-wildlife/fungi-and-lichens/beefsteak-fungus/#:~:text=Beefsteak%20fungus%20is%20common%20in,sought%20after%20by%20furniture%20makers
30 Luis Villazon, 'Am I more bacteria than human?' sciencefocus.com/the-human-body/am-i-more-bacteria-than-human/
31 Peter Wohlleben, *The Hidden Life of Trees* (Vancouver: Greystone Books Ltd, 2014), 114.
32 Brian Douglas and Kay Haw, 'The Lost and Found Fungi Project', *Wood Wise – Fabulous Fungi* (November 2015), woodlandtrust.org.uk/publications/2015/11/wood-wise-fabulous-fungi/ page 15

33 'The wonders of fungi', *Wood Wise – Fabulous Fungi* (Autumn 2015), woodlandtrust.org.uk/publications/2015/11/wood-wise-fabulous-fungi/ page 3
34 Woodland Trust, 'Ancient Woodland', woodlandtrust.org.uk/trees-woods-and-wildlife/habitats/ancient-woodland/
35 Geoff Frampton and Steve Hopkin, 'Springtails – in search of Britain's most abundant insects', 6 August 2001, britishwildlife.com/article/article-volume-12-number-6-page-402-410/
36 George Monbiot, 'The secret world beneath our feet is mind-blowing – and the key to our planet's future', *Guardian*, 7 May 22, theguardian.com/environment/2022/may/07/secret-world-beneath-our-feet-mind-blowing-key-to-planets-future

7: THREATS: COULD A BEAUTY BE A BEAST?

1 Action Oak, 'About', actionoak.org/about
2 Forest Research, 'Tools and Resources, Xylella (Xylella fastidiosa)', forestresearch.gov.uk/tools-and-resources/fthr/pest-and-disease-resources/xylella-xylella-fastidiosa/
3 Rackham, *Woodlands*, 319–320
4 Forest Research, 'Two-spotted oak buprestid (Agrilus biguttatus)', forestresearch.gov.uk/tools-and-resources/fthr/pest-and-disease-resources/two-spotted-oak-buprestid-agrilus-biguttatus/
5 Sandra Denman, 'Updating our understanding of Acute Oak Decline', Bacterial Plant Diseases Programme, 16 February 2022, bacterialplantdiseases.uk/updating-our-understanding-of-acute-oak-decline/
6 Lorienne Whittle, ' "From little acorns mighty oaks grow" – Why we should be worried for our native oak trees', Woodland Trust blog, 9 November 2021, naturescalendar.woodlandtrust.org.uk/blog/2021/oak-masting-2021/
7 In most cases a trip to your local chemist will help: Public Health England, 'Health effects of exposure to setae of oak processionary moth larvae, Systematic review', assets.publishing.service.gov.uk/government/uploads/system/uploads/attachment_data/file/432003/Oak_Processionary_Moth_FINAL__2_.pdf page 6
8 C. Beverley, 'Thaumetopoea processionea (oak processionary moth)', March 2014, cabidigitallibrary.org/doi/10.1079/cabicompendium.53502
9 Forest Research, 'Tools and Resources Oak processionary moth (Thaumetopoea processionea)', forestresearch.gov.uk/tools-and-

resources/fthr/pest-and-disease-resources/oak-processionary-moth-thaumetopoea-processionea/
10 Woodland Trust, 'State of the UK's Woods and Trees 2021', page 147
11 Woodland Trust, 'Tree Pests and Diseases', woodlandtrust.org.uk/trees-woods-and-wildlife/tree-pests-and-diseases/
12 Forest Research, 'Dutch elm disease: History of the Disease', forestresearch.gov.uk/tools-and-resources/fthr/pest-and-disease-resources/dutch-elm-disease-ophiostoma-novo-ulmi/dutch-elm-disease-history-of-the-disease/
13 Woodland Trust, 'Tree Pests and Diseases'
14 Dr Ruth Mitchell, James Hutton Institute, 'Revealing the hidden impact of plant pests and pathogens', 24 August 2022, stories.rbge.org.uk/archives/36592#
15 Woodland Trust, 'Ancient Woodland', woodlandtrust.org.uk/trees-woods-and-wildlife/habitats/ancient-woodland/
16 Philip Hibble, 'Another day, another ancient tree felled: 300-year-old Hunningham Oak near Leamington is knocked down to make way for HS2', *Northampton Chronicle*, 28 Sep 2020, northamptonchron.co.uk/news/environment/another-day-another-ancient-tree-felled-300-year-old-hunningham-oak-near-leamington-is-knocked-down-to-make-way-for-hs2-2984031
17 Woodland Trust, 'HS2 Rail Link', woodlandtrust.org.uk/protecting-trees-and-woods/campaign-with-us/hs2-rail-link/
18 'Sir David Attenborough joins Instagram to warn "the world is in trouble"', 24 September 2020, https://www.bbc.co.uk/news/entertainment-arts-54281171
19 Woodland Trust, 'Campaign with us', woodlandtrust.org.uk/protecting-trees-and-woods/campaign-with-us/
20 Forestry Research, 'Forest carbon stock', forestresearch.gov.uk/tools-and-resources/statistics/forestry-statistics/forestry-statistics-2018/uk-forests-and-climate-change-5/forest-carbon-stock/
21 Lewis-Stempel, *The Glorious Life of the Oak*, 23
22 Met Office, 'New Temperature Records for the UK', metoffice.gov.uk/about-us/press-office/news/weather-and-climate/2022/new-years-day
23 Woodland Trust, 'Nature's Calendar', naturescalendar.woodlandtrust.org.uk/
24 Woodland Trust, 'State of the UK's Woods and Trees 2021', page 125
25 Woodland Trust, 'State of the UK's Woods and Trees 2021', page 127

26 Woodland Trust, 'Tackling air pollution with trees', woodland trust.org.uk/trees-woods-and-wildlife/british-trees/tackling-air-pollution-with-trees/

27 British Ecological Society, 'We Need to Talk About Nitrogen', 13 March 2017, britishecologicalsociety.org/need-talk-nitrogen

28 Woodland Trust, 'Deer', January 2020, woodlandtrust.org.uk/media/47750/deer-position-statement.pdf page 2

29 David Oakes, 'The Pinaceae', *Trees a Crowd*, 20 April 2021, treesacrowd.fm/56trees/

31 RSPCA, 'Road Traffic Accidents Involving Deer', 2015, science.rspca.org.uk/documents/1494935/9042554/Roadtraffic+accidents+involving+deer+%28v1.0%29+-+2015.pdf/ac2037be-5fc0-ff47-7834-092c6ba51325?t=1553171460915#:~:text=This%20is%20a%20major%20animal,or%20swerving%20to%20avoid%20deer page 1

31 Will Richardson, Dr Glyn Jones, Martin Glynn, Peter Watson, 'An Analysis of the Cost of Grey Squirrel Damage to Woodland', Royal Forestry Society, 2021, rfs.org.uk/wp-content/uploads/2021/03/analysis-of-the-cost-of-grey-squirrel-damage-to-woodland-publication-copy-180121.pdf

32 Met Office, 'Lessons and legacy of the Great Storm of 1987', metoffice.gov.uk/about-us/who/our-history/lessons-and-legacy-of-the-great-storm-of-1987

33 BBC Newsround, ' "Blah, blah, blah" Greta Thunberg criticises UK government at Youth4Climate', 29 September 2021, bbc.co.uk/newsround/58731698

34 Woodland Trust, 'State of the UK's Woods and Trees 2021', page 16

35 Europe Economics, 'The economic benefits of woodland: A report for the Woodland Trust prepared by Europe Economics', January 2017, woodlandtrust.org.uk/publications/2017/01/economic-benefits-of-woodland page 3

36 Jessica Nolan, '49 Types of Oak Trees (with Pictures): Identification Guide', leafyplace.com/oak-tree-types-bark-leaves/

8: THE OAK'S LITTLE HELPERS. AND ITS BIG HELPERS.

1 Fritha West, 'Is 2022 a mast year?', Woodland Trust blog, 26 October 2022, naturescalendar.woodlandtrust.org.uk/blog/2022/is-2022-a-mast-year/

2 Jamie Wyver, 'Photo of the week: the joy of jays', 7 September 2020, community.rspb.org.uk/ourwork/b/natureshomemagazine/posts/photo-of-the-week-young-jay-with-parent

3 British Trust for Ornithology, 'Jay Garrulus glandarius', bto.org/our-science/projects/gbw/gardens-wildlife/garden-birds/a-z-garden-birds/jay
4 Ancient Tree Forum, 'Bellmount Ancient Burry Oak, Belton Park, Lincolnshire', ancienttreeforum.org.uk/ancient-trees/ancient-tree-sites-to-visit/east-of-england/bellmount-ancient-burry-oak-belton-park/
5 University of Cambridge, 'Monogamous birds read partner's food desires', 15 February 2013, cam.ac.uk/research/news/monogamous-birds-read-partners-food-desires
6 Ian M, 'Five facts about jays', RSPB, 27 September 2021, community.rspb.org.uk/ourwork/b/scotland/posts/five-facts-about-jays
7 RSPB, 'Jay', rspb.org.uk/birds-and-wildlife/wildlife-guides/bird-a-z/jay/
8 Miles, The British Oak, 24
9 Museum Blogger, 'The Mary Rose – Racing Yacht?', 17 March 2015, maryrose.org/blog/historical/museum-blogger/the-mary-rose-racing-yacht/
    See also Medieval Histories, 'The Mary Rose', 4 June 2013, medieval.eu/a-brand-new-museum-shows-the-splendors-of-the-mary-rose-the-flagship-of-henry-viii/
10 Royal Navy, 'Scottish timber will help HMS Victory's restoration', 15 February 2016, royalnavy.mod.uk/news-and-latest-activity/news/2016/february/15/160215-scottish-timber-hms-victory
11 Simon Schama, 'The tree that shaped Britain', 7 May 2010, news.bbc.co.uk/1/hi/8668587.stm
12 Lewis-Stempel, The Glorious Life of the Oak, 22
13 Hainault Forest, 'Hainault Forest has had a very turbulent history', hainaultforest.org/about/history/
14 Woodland Trust, 'Emergency Tree Plan for the UK. How to increase tree cover and address the nature and climate emergency', January 2020, woodlandtrust.org.uk/media/47692/emergency-tree-plan.pdf
15 House of Commons Library, 'Tree Planting in the UK', 2 June 2021, researchbriefings.files.parliament.uk/documents/CBP-9084/CBP-9084.pdf page 7
16 Forest Research, 'Forestry Statistics and Forestry Facts & Figures', 28 September 2023, forestresearch.gov.uk/tools-and-resources/statistics/forestry-statistics/
17 IPCC, 'Special Report: Global Warming of 1.5°C', ipcc.ch/sr15/chapter/chapter-1/
18 knepp.co.uk

19 Charlie Burrell, 'Forging a new path', *Wood Wise*, Autumn 2020, woodlandtrust.org.uk/media/49178/woodwise-woods-in-waiting-autumn-2020.pdf page 5
20 Charlie Burrell, 'Forging a new path', 5
21 Joan Cottrell, 'Natural regeneration promotes genetic adaptation', *Wood Wise*, Autumn 2020, woodlandtrust.org.uk/media/49178/woodwise-woods-in-waiting-autumn-2020.pdf page 19
22 Abi Bunker, 'Natural regeneration in dynamic woods and Landscapes', *Wood Wise*, Autumn 2020, woodlandtrust.org.uk/media/49178/woodwise-woods-in-waiting-autumn-2020.pdf page 3
23 Woodland Trust, 'Campaign with Us'
24 David Hirst, 'Public forest estate sell-off – what next?', 8 December 2014, commonslibrary.parliament.uk/public-forest-estate-sell-off-what-next/
25 'The Queen and Prince Charles "Plant a Tree for the Jubilee" to Kick-start Green Canopy Project', The Royal Family Channel [unofficial], youtube.com/watch?v=Zq75nIU99RI

SOME FUN STUFF
1 Plantlife, 'No Mow May', plantlife.org.uk/campaigns/nomowmay/
2 RSPB, 'Nature's Calendar: May', rspb.org.uk/birds-and-wildlife/natures-calendar/natures-calendar-may#dawn-chorus-day
3 See naturescalendar.woodlandtrust.org.uk
4 See wildlifetrusts.org/30-days-wild
5 See bigbutterflycount.butterfly-conservation.org

Select Bibliography

Baker, Margaret, *Discovering the Folklore of Plants* (Oxford: Shire Publications, Bloomsbury Publishing, 2008).
Binney, Ruth, *Plant Lore and Legend* (Hassocks, West Sussex: Rydon Publishing, 2016).
Brock, Paul D., *Britain's Insects: A Field Guide to the Insects of Great Britain and Ireland* (New Jersey/Woodstock: Princeton University Press, 2021).
Canton, James, *The Oak Papers* (Edinburgh: Canongate Books, 2020).
Dalton, Stephen and Jill Bailey, *The Secret Life of an Oakwood* (London: Century Hutchinson, 1986).
Dobson, Frank S., *Lichens: An Illustrated Guide to the British and Irish Species* (London: the British Lichen Society in collaboration with the Richmond Publishing Co., 2018).
Farjon, Aljos, *Ancient Oaks in the English Landscape* (London: Kew Publishing, 2017).
Frazer, Sir James George, *The Golden Bough: A Study of Magic and Religion* (Mineola, NY: Dover Books, 2003. First published in 1906 and 1915, abridged in 1922).
Gerard, John, *Gerard's Herbal* (London: Senate, 1994. First published in 1597).
Gilbert, Oliver, *Lichens, Collins New Naturalist series* (London: HarperCollins, 2000).
Gorman, Gerard, *Woodpeckers, RSPB Spotlight* (London: Bloomsbury Wildlife, 2018).
Harding, Paul T. and Tom Wall, eds, *Moccas: An English Deer Park* (English Nature, 2000).
Harford, Robin, *Oak Notebook* (Great Britain: Amazon, 2018).
Hight, Julian, *Britain's Ancient Forest: Legacy & Lore* (Great Britain: Julian Hight, 2019)
Hight, Julian, *Britain's Tree Story: The history and legends of Britain's Ancient Trees* (London: National Trust Books, 2011).

Lewis-Stempel, John, *The Glorious Life of the Oak* (London: Doubleday Publishers, 2018).
Logan, William Bryant, *Oak: The Frame of Civilization* (New York: W. W. Norton and Company, 2006).
Mabey, Richard, *Food for Free* (London: HarperCollins, 2007).
Mabey, Richard, *Plants with a Purpose* (Glasgow: Fontana/Collins, 1977).
Marren, Peter, *Britain's Ancient Woodland Heritage* (Devon: David and Charles Publishers, 1990).
McMorland Hunter, Jane, ed., *A Nature Poem for Every Day of the Year* (London: Portico, 2018).
Miles, Archie, *The British Oak* (London: Constable, 2016).
Morris, M. G. and F. H. Perring, eds, *The British Oak: Its History and Natural History* (Berkshire: W. E. Classey, 1974).
Nozedar, Adele, *The Tree Forager* (Cornwall: TJ Books, 2021).
Ó'Nualláin, Fiann, *First Aid from the Garden* (Cork: Mercier Press, 2016).
Ó'Nualláin, Fiann, *The Holistic Gardener: Natural Cures for Common Ailments* (Cork: Mercier Press, 2016).
Porley, Ron and Nick Hodgetts, *Mosses and Liverworts, The New Naturalist Library* (London: Collins, 2005).
Rackham, Oliver, *Woodlands* (London: William Collins, 2015).
Redfern, Margaret and Peter Shirley, *British Plant Galls* (Shrewsbury: FSC Publications, 2011).
Schneidau, Lisa, *Botanical Folk Tales of Britain and Ireland* (Cheltenham: The History Press, 2018).
Southwood, T. R. E., *Life of the Wayside & Woodland: A Descriptive Seasonal Guide* (London: Frederick Warne and Co., 1963).
Tait, Malcolm, ed., *Wildlife Gardening for Everyone* (London: Think Books, 2007).
Tree, Isabella, *Wilding* (London: Picador, 2018).
Wohlleben, Peter, *The Hidden Life of Trees* (Vancouver: Greystone Books, 2014).
Wohlleben, Peter, *The Power of Trees* (Vancouver: Greystone Books, 2014).
Wright, John, *The Forager's Calendar: A Seasonal Guide to Nature's Wild Harvests* (Cumbria: Profile Books, 2020).
Young, Francis Brett, *The Island* (Kingswood: William Heinemann, 1944).

Index

A

acid rain, 124
acorns
    acorn flour bread roll recipe, 136, 158–59
    balanophagy (acorn-eating), 134–38
    growing season, 225
    health claims, 142
    jays and, 189–90
    magic and, 114
    mast years, 186–87, 226
    pedunculate v. sessile oaks, 9–10
acorn weevil, 135
Action Oak, 161, 162, 208
activism, 181, 182–83, 207, 214. *See also* conservation
Acton (surname), 58
acute oak decline (AOD), 161, 163–65
age, of trees, 57
air pollution, 124, 175
algae, 122
Amanita fungi (grisettes), 145–46
Amateur Entomologists' Society, 43
Ancient Burry Oak, 190
Ancient Tree Inventory (ATI), 11–12, 35, 104, 208, 213
ancient trees
    characteristics, 8, 11–12
    Charles III on, 208
    discoveries, ongoing, 104, 208, 213
    generation gap, 104–5
    legal protections, 181, 205–6
    veteranisation, 105–6
Andricus quercuscalicis, 45
ants, 64, 68, 73
aphids, 72–73
ash dieback, 166–67
ash trees, 118–19, 154, 166–67, 190, 194
Attenborough, David, 169
Auden, W. H., 98
autumn, 226

B

bacteria, 152, 164
Baker, Margaret, 114
balanophagy (acorn-eating), 134–38
bark beetles, 106
Barkham, Patrick, 73
bats, 15, 106, 226
BBC Radio 4
    Tweet of the Day, 27, 29
beard (*Usnea*) lichens, 123, 124
Beardshaw, Chris, 219
Bechstein's bat, 15
beech trees, 154, 187
beefsteak fungus (*Fistulina hepatica*), 133, 146, 148
beetles
    bark beetles, 106
    blue ground beetle, 97
    lesser stag beetle, 96
    Moccas beetle, 97–98

oak jewel beetle, 161, 163–65
saproxylic beetles, 91–92, 102, 103
spotted longhorn (black-and-yellow longhorn), 92–94, 95
The Bee Tree, 31, 34
Belton House (Lincolnshire), 188–89, 190
The Big Belly Oak, 115–16
Big Butterfly Count, 74, 214, 225
Big Garden Birdwatch, 214
biodiversity, 42, 176, 203
Biorhiza pallida, 43–44
birch trees, 94, 118–19, 154, 190
birds
    Big Garden Birdwatch, 214
    bird feeders, 216, 224
    caterpillars and, 22, 23–24, 68, 72, 174, 193
    climate change and, 174
    drumming and other noises, 29–30
    International Dawn Chorus Day, 224
    lichens and, 127–28
    sexes, differences between, 21
    songs, listening to and identifying, 25–28, 28–29
    wildlife-friendly spaces for, 72
birds, specific species
    blackbird, 28–29
    blue tit, 14, 22, 23–24, 47, 72, 174
    bumbarrel, 127–28
    chiffchaff, 29, 38, 224
    coal tit, 22
    dunnock, 29
    great spotted woodpecker, 14, 16, 17, 18, 19, 25, 30
    great tit, 14, 22, 174
    green woodpecker, 17, 19, 30
    jay, 189–91, 226
    lesser spotted woodpecker, 17, 18, 19, 25, 30
    long-tailed tit, 14, 26, 127
    mistle thrush, 28–29
    nuthatch, 22, 35
    pied flycatcher, 20–22, 23–24, 174, 226
    redstart, 22
    robin, 27, 28, 47, 173
    song thrush, 27–28, 28–29
    starling, 21–22
    tawny owl, 226
    woodpecker, 14–18, 19–20, 24, 25, 30, 47, 106, 224
    wren, 29
    wryneck, 18–19
The Birnam Oak, 54–55
The Birnam Sycamore, 54
black-and-yellow longhorn beetle (spotted longhorn, *Rutpela maculata*), 92–94, 95
blackbird, 28–29
blackthorn, 193
Blanchard, Richard, 49–50
bluebells, 215, 225
blue ground beetle (*Carabus intricatus*), 97
blue tit, 14, 22, 23–24, 47, 72, 174
Boddy, Lynne, 148, 155
bolete mushrooms, 145
Boniface (saint), 113
boring insects, 107
Boscobel House (Shropshire), 78, 80
The Boscobel Oak (The Royal Oak), 78–79, 80
Boutcher, William, 76
The Bowthorpe Oak, 48–51, 147
branding and marketing, 58–59
Branscombe, Julian, 27–28, 223
Brenneria goodwinii, 164
British Trust for Ornithology (BTO), 18–19, 189–90, 206, 214
Broadwells Wood (Warwickshire), 170

Brock, Paul D.
   Britain's Insects, 93
The Bruce Tree (The Strathlevan House Oak), 83
Brussels lace moth, 128
Bryce, Marion, 130
Bryn Arw (Brecon Beacons), 20–21
bryophytes, 3, 118–19, 119–21, 130, 223. *See also* lichens
Buglife, 97, 206, 214, 225
bumbarrel, 127–28
Bunker, Abi, 204
Burgess, Malcolm, 72
Burrell, Charlie, 203, 209
butterflies
   Big Butterfly Count, 74, 214, 225
   hunting tips, 74
   purple hairstreak butterfly, 62–65, 65–66, 74
   See also caterpillars; moths
Butterfly Conservation, 64, 67, 68, 76, 206, 214, 225
The Buttington Oak, 104
Buxton, Edward North, 195, 207

C

caddisflies, 69
Caesar's mushrooms (*Amanita caesarea*), 146
Canton, James, 5, 58, 107
carbon
   footprint, 215
   net zero carbon emission goals, 181, 196–97
   storage, 108, 157, 171, 205
Careless, William, 78–79
casks, 139–40
caterpillars, 22, 23–24, 64, 67–68, 72, 128, 165–66, 174, 193
cats, 216, 225
Celtic rainforest, 117–19
ceps (penny buns, porcini, *Boletus edulis*), 145

Chapman, Emma, 143–44
Charles II (king), 78–79
Charles III (king), 77, 80, 169, 208
Charter for Trees, Woods and People (Tree Charter), 46–47
Charter of the Forest, 47
chicken of the woods (*Laetiporus sulphureus*), 142–43, 146
chiffchaff, 29, 38, 224
Chinery, Michael
   Britain's Plant Galls, 46
Christianity, 112, 113–14
citizen science, 213–14
The Clachan Oak, 117
Clare, John
   'Burthorp Oak', 48–49
   'Remembrances', 171
Clark, Eleanor, 99
Clarke, Danny, 199, 216–17, 218–19
Claudius (Roman emperor), 146
Clean Air Acts, 124, 175, 182–83
climate change, 155, 157, 164, 173–74, 196–97, 204
coal tit, 22
Cobbett, William, 137
Coed Maesmelin, 97
common emerald moth, 71
common oak. *See* pedunculate oak
common rhododendron (*Rhododendron ponticum*), 167
compaction, soil, 179–80
conservation
   activism, 181, 182–83, 207, 214
   charities, 206, 214
   charity staff and volunteers, 206, 208, 214–15
   Charles III and, 208
   false promises, 181–82
   by farmers and landowners, 209
   by 'gods of small things', 209

history of, 195
hope for, 197–98
individual responsibility for, 211–12
legal protections, 181, 205–6
planting v. natural regeneration, 202–4
Young People's Forest, 65, 198–200, 200–202, 207
See also threats, to oaks
consumer power, 215
Cornish, Carl, 15, 19, 21, 99
Cornish oak. *See* sessile oak
Cottrell, Joan, 204
Couzens, Dominic, 29
Cowdray Park (Midhurst, West Sussex), 84–85
cow parsley (Queen Anne's Lace), 93
Cowper, William
'Yardley Oak', 57, 109
Craven, John, 122
crotal (crottle, *Parmelia sacatilis*) lichen, 123–24
The Crouch Oak, 85
The Curley Oak, 208

D

Darwin, Charles, 19, 97
da Vinci, Leonardo, 16, 20
dead wood, 89–91, 91–92, 96–97
deathcap (*Amanita phalloides*), 146
deer, 176–77
deer parks, 102–3
development, as threat, 169–71
devil, summoning, 115–16
diseases and pests, trees, 166–67, 214
Dobson, Frank S., 123
Dodona (Greece), 112
dog rose, 193
dogs, 147, 216, 221, 225
dreams, 115
Druids, 110–11
Dryden, John, 56

dunnock, 29
durmast oak. *See* sessile oak

E

Eastern Claylands (Suffolk and Essex), 104
edibles
    acorn flour bread roll recipe, 136, 158–59
    acorns, 134–38
    beefsteak fungus, 133, 146
    boletes, 145
    chicken of the woods, 146
    fungi, 134, 145–47
    grisettes (Amanita fungi), 145–46
    health warning, 134
    hen of the woods, 146
    oak apples, 139
    oak leaves, 138
    oak wood, for smoking, 138
    summer truffle, 147
    whisky, aged in casks, 139–40
Edwards, Sean R., 120
Edward the Confessor, 77
Edward VI (king), 141
elephant hawk-moth, 70
Elizabeth I (queen), 77, 83–86, 193
Elizabeth II (queen), 77, 84, 208
elm trees, 166
Emerson, Ralph Waldo, 210
Enclosure Acts, 171
English Heritage, 80
English oak. *See* pedunculate oak
Epping Forest, 26, 195, 207
European oak. *See* pedunculate oak
Evelyn, John, 142
evergreen oak (holm oak, holly oak, *Quercus ilex*), 8, 137
Eversham, Brian, 4, 125–27, 129–30, 131, 224
executions, 117

F

fairies, 110, 112, 120
Falk, Steven, 94
Farjon, Aljos, 5
Fellowes, Lord and Lady, 209
Fingle Woods (Dartmoor), 143
Firth, Alasdair, 118
folklore, 110, 124. *See also* fairies; magic and witchcraft; medicines and health claims
food. *See* edibles
Forbes Hole Nature Reserve, 68–71
forest bathing, 143
Forest Research, 165
forests and woods, 97, 98–102, 143–45, 156. *See also* trees
Frazer, James George, 111–12, 113
Frith, Mark, 51
Froglife, 206
Fuller, Isaac, 79
fungi
  about, 3, 154–55
  edible, 134, 145–47
  galls and, 41
  heart wood, consumption by, 147–49
  lichen and, 122
  mycorrhizal fungi, 150, 153, 154, 156
  predators of, 157
  soil health and, 155–57
  symbiotic relationship with trees, 149–52, 152–54, 155, 156, 193
  veteranisation and, 106
fungi, specific species
  beefsteak, 133, 146, 148
  boletes, 145
  chicken of the woods, 142–43, 146
  grisettes (Amanita fungi), 145–46
  hen of the woods, 146
  penny buns (ceps, porcini), 145
  summer bolete, 145, 149–50
  summer truffle, 147

G

gall midges, 42
galls
  author's experience, 43, 44
  guests and parasites within, 40–41
  hedgehog galls, 45
  identification tips, 46
  ink from, 37, 38–39, 46
  knopper galls, 45
  on lichens, 128–29
  Macrodiplosis pustularis galls, 42
  oak apples, 43–44, 115, 139
  oak marble galls, 37, 38, 39, 40–41, 45, 47, 141
  oaks and, 42
  searching for (galling), 44–46, 225
  silk button galls, 45–46, 88
  spangle galls, 44–45
  Trioza remota galls, 42–43
  within web of life, 47
gall wasps
  Andricus quercuscalicis, 45
  Biorhiza pallida, 44
  hedgehog gall wasp, 45
  marble gall wasp, 37, 38, 39, 40–41, 59
  silk button gall wasp, 45–46
  spangle gall wasp, 45
gardening, wildlife, 72, 96, 218–19
Gardner, Charlie, 182
gender, 86
generation gap, trees, 104–5
ghost woods, 100
giant hogweed (*Heracleum mantegazzianum*), 95

Gilbert, Oliver, 122, 125, 128
Glyndwr, Owain, 82
'gods of small things', 107–8, 209
Gorman, Gerard, 18, 19, 30
  Woodpeckers, 17
Gosling, Rebecca, 161, 164
gospel oaks, 113–14
great spotted woodpecker (*Dendrocopos major*), 14, 16, 17, 18, 19, 25, 30
great tit, 14, 22, 174
Greece, 112
green oak tortrix moth (*Tortrix viridana*), 23, 67–68
green woodpecker (*Picus viridis*), 17, 19, 30
grey squirrel, 177, 178, 192
grisettes (Amanita fungi), 145–46

H

habitat translocation, 170–71
Hackett, Louise, 88, 89, 90–91, 96, 98–99, 100, 101, 106–7, 172, 218
Hainault Forest (Essex), 194–96, 207
Haldane, J. B. S., 87
hammered shield lichen (*Parmelia sulcata*), 128
Hampton, Helen, 224
The Hanging Tree, 117
Hardy, Thomas
  'An August Midnight', 216
  Far from the Madding Crowd, 58
Harford, Robin, 114
hawk-moths, 70
Hawksworth, David, 122
hawthorn, 193
Heal, James, 36, 40, 41, 42, 43, 44, 45–46
health claims and medicines, 47, 123, 141–45, 202
heart wood, 90–91, 147–48

hedgehog galls, 45
hedgehog gall wasp (*Andricus lucidus*), 45
hedgerows, 157, 172, 219
Hemiptera (true bugs), 43
hen of the woods (*Grifola frondosa*), 146
Henry IV (king), 82
The Herne's Oak, 52–53
Herne the Hunter, 52–53, 103
Hesiod, 135
Hight, Julian, 5, 83, 84
  Britain's Tree Story, 78
Hodgetts, Nick, 119, 120, 121
Hodgson, Julian, 73
hogweed (*Heracleum sphondylium*), 93, 95
holm oak (evergreen oak, holly oak, *Quercus ilex*), 8, 137
honeydew, 73
hornworts, 119, 121. *See also* bryophytes
Howitt, Mary, 61
HS2 (high-speed railway), 169–70
Hughes-Hallett, James and Alka, 37–38, 209
humans
  'gods of small things', 107–8, 209
  oaks and, 30–34, 193–98
  See also conservation
The Hunningham Oak, 169
Hussein, Sabaha, 202, 207
hybrid oaks (*Quercus x rosacea*), 10
Hyde, Ruth, 208

I

Ickworth Park (Suffolk), 33
immortality, 32, 149
imports, 166
ink, oak gall (iron gall ink), 37, 38–39, 46

insect inquilines, 40
insects, 59–60, 215–16. *See also* beetles; butterflies; gall wasps; invertebrates; moths; spiders
Institute of Chartered Foresters (ICF), 47
Intergovernmental Panel on Climate Change (IPCC), 197
International Dawn Chorus Day, 224
invasive (non-native) plants, 167–68
invertebrates, 3, 23, 59, 90–91, 92, 94, 219. *See also* insects
Ireland, Samuel Herne's Oak, Windsor Park, 53
Irish oak. *See* sessile oak
iron prominent moth, 71
ivy, 226

J

James Hutton Institute, 3
Janega, Eleanor, 113
jay (*Garrulus glandarius*), 189–91, 226
John (king), 33, 77
Johnes, Thomas, 194
Jukes, Jonathan, 195
jumping plant lice (psyllid bugs), 42

K

Kedleston Park, 63
Kelvingrove Park (Glasgow), 32
Kettlewell, Bernard, 71
Kett's Oaks, 33–34
Kew Gardens, 179
Khan, Amir, 144–45
King, Simon, 9
King Offa's Oak, 102–3
Kirkham, Tony, 56, 179–80, 181, 184, 217–18
Knepp Wildland Project (West Sussex), 173, 203, 209

knopper galls, 45

L

The Lady-in-Waiting Oak, 85
Lammas Day (Loaf Mass Day), 24
Lammas leaves, 24, 68, 186
Lancashire Wildlife Trust, 75
language, 58
lawn mowing, 219
Leafy Place (website), 184
leaves, 10, 24, 68, 138, 183–84, 186
legal protections, 181, 205–6
lesser spotted woodpecker (*Dryobates minor*, *Dendrocopos minor*), 17, 18, 19, 25, 30
lesser stag beetle (*Dorcus parallelipipedus*), 96
Lewis, Amy, 18
Lewis-Stempel, John, 5, 25, 76–77
lichens, 3, 118–19, 121–25, 127–29, 130, 223. *See also* bryophytes
light emerald moth, 71
lightning, 115
literature
   'Burthorp Oak' (Clare), 48–49
   Far from the Madding Crowd (Hardy), 58
   Georgics (Virgil), 56, 57
   ink, from galls, 37, 38–39, 46
   The Lord of the Rings (Tolkien), 55
   oaks, inspiration from, 47–48
   'Remembrances' (Clare), 171
   Shakespeare, 6, 40, 51–52, 54, 55, 100
   'Yardley Oak' (Cowper), 57, 109
The Little Owl Oak, 146
Littlewood, Nick, 65, 66
liverworts, 119, 121. *See also* bryophytes
Living Landscapes, 101, 173

Lloyd George, David, 139
Logan, William Bryant, 5, 38–39, 135, 137
log piles, 96–97, 218
London, 7
London Natural History Society (LNHS), 40
Long Eaton Natural History Society, 68, 214
long-tailed tit, 14, 26, 127
love, 5–6, 186, 213
lynx, 176

M

Mabey, Richard, 138
MacLean, Charles, 140
Macrodiplosis pustularis galls, 42
Maggie née Ta-bu-ce, 135–36
magic and witchcraft, 114–15, 126
The Major Oak, 31–32, 98, 102, 180
The Man-eating Caterpillar Tree, 34–35
marble gall wasp (*Andricus kollari*), 37, 38, 39, 40–41, 59
marketing and branding, 58–59
The Marton Oak, 33, 133–34, 141, 147, 149
mast years, 186–87, 226
McGavin, George, 39, 59
Mead, Margaret, 211
medicines and health claims, 47, 123, 141–45, 202
The Medusa Oak, 35, 88, 89–90, 91
mental health, 143–45, 202
Miles, Archie, 5, 33, 50, 58, 82, 83, 114, 193
  The British Oak, 35, 78
mistle thrush, 28–29
mistletoe, 111
Mitchell, Karl, 189
Mitchell, Ruth, 3, 4
Moaning Mabel (ancient oak), 34

Moccas beetle (*Hypebaeus flavipes*), 97–98
Monbiot, George, 157
Monks Wood (Cambridgeshire), 73
Moorcroft, Darren, 1, 8, 59, 196
mosses, 119–20, 121. *See also* bryophytes
moths
  about, 66–67
  lichen and, 128
  Moth Night experience, 68–71
  moth spotting, tips for, 75–76
  See also butterflies; caterpillars
moths, specific species
  Brussels lace moth, 128
  common emerald moth, 71
  green oak tortrix moth, 23, 67–68
  hawk-moths, 70
  iron prominent moth, 71
  light emerald moth, 71
  mottled umber, 23, 67
  oak processionary moth, 165–67
  peppered moth, 71, 128
  The Prominents, 75–76
  winter moth, 23, 67
mottled umber (*Erannis defoliaria*), 23, 67
Mulholland, Jim, 149, 180
muntjacs, 116
mushrooms. *See* fungi

N

National Lottery Heritage Fund, 143
National Planning Policy Framework (NPPF), 206
National Trust, 59, 63, 143, 206, 214
nature
  forests and woods, 97, 98–102, 143–45, 156

interconnectedness, 72, 192–93
natural regeneration, 202–4
reconnecting landscapes, 172–73
seasonal delights, 223–26
tips for enjoying, 220–23
UK's woodland cover targets, 181, 196–97
wildlife-friendly spaces, caring for, 72, 94–95, 96, 218–19
See also conservation; trees
Nature's Calendar, 173, 213, 225
New Forest, 100–101, 125–27, 129–30
Nichols, Chris, 178, 198
nitrogen pollution, 124–25, 175
Nolan, Victoria, 104
non-native (invasive) plants, 167–68
Normans, 102–3
Northern Forest, 101
notable trees, 12
Nozedar, Adele
The Tree Forager, 136
nuthatch (*Sitta europaea*), 21, 35

O

Oak Apple Day (Royal Oak Day), 80–81
oak apples, 43–44, 115, 139
Oakes, David, 176–77
oak gall ink (iron gall ink), 37, 38–39, 46
oak jewel beetle (*Agrilus biguttatus*), 161, 163–65
oak marble galls, 37, 38, 39, 40–41, 45, 47, 141
oak moss (*Evernia prunastri*), 119, 125–26, 131, 223
The Oak of Reformation, 33–34
oak processionary moth (OPM, *Thaumetopoea processionea*), 165–67

oaks
acorns (*see* acorns)
ancient oaks, 8, 10–12
animal helpers, 188–92
appearance and identification, 8–10
branding and marketing, 58–59
conservation (*see* conservation)
edibles (*see* edibles)
extraordinariness of, 2–3
fungi, symbiotic relationship with, 149–52, 152–54, 155, 193
galls (*see* galls)
gender, 86
growing, tips for, 216–17
habitat, 49, 98, 102–3
health claims, 47, 141–45
hollowing out, 147–49
humans and, 30–34, 193–98
immortality, 32, 149
imported trees, 166
language and, 58
leaves, 10, 24, 68, 138, 183–84, 186
literature and (*see* literature)
love for, 5–6, 186, 213
mast years, 186–87, 226
recommended books, 5
reproduction, 187–88
roots, 56–57, 150–52, 180
royalty and (*see* royalty, and oaks)
species associated with, 3–4, 20, 23, 59
spirituality (*see* spirituality)
as survivors, 162
threats to (*see* threats, to oaks)
in UK, 7–8
oaks, specific species
holm oak, 8, 137
hybrids, 10
pedunculate oak, 7, 9–10, 45
red oak, 8
sessile oak, 7, 9–10, 33

Turkey oak, 8, 39, 45
white oak, 137
Observatree, 213
One in a Million leaf activity, 183–84
Ó'Nualláin, Fiann, 142
Orpe, Ken, 62, 63, 64–65, 201
Ötzi, 120
The Owen Glendower Oak, 33, 82–83

P

pale giant oak aphid, 73
Pankhurst, Emmeline, 32
pannage, 137–38
The Panshanger Oak, 86
parasitoids, 40–41
The Parliament Oak, 33, 77, 103, 149
pedunculate oak (English oak, common oak, European oak, *Quercus robur*), 7, 9–10, 45
Penn, Robert, 21
penny buns (ceps, porcini, *Boletus edulis*), 145
peppered moth, 71, 128
Peppered Moth Game, 71–72
Pepys, Samuel, 80–81
Percy, Henry 'Hotspur', 82
pests and diseases, trees, 166–67, 214
Phoma gallorum, 41
pied flycatcher (*Ficedula hypoleuca*), 20–22, 23–24, 174, 226
pigs, 137–38
pine marten, 192
Plantlife, 124–25, 130, 175, 206, 214, 219
Plant Parasites of Europe (website), 46
pollarding, 12, 86
pollution, 71, 124–25, 128, 174–75
ponds, 219
poplar hawk-moth, 70, 71

porcini (penny buns, ceps, *Boletus edulis*), 145
Porley, Ron, 119, 120, 121
Powell, Keith, 21
Powell, Mark, 131
privet hawk-moth, 70
The Prominents (moths), 75–76
psyllid bugs (jumping plant lice), 42
pubs, 79–80, 82
purple hairstreak butterfly (*Favonius quercus*), 62–65, 65–66, 74

Q

Queen Anne's Lace (cow parsley), 93
The Queen Elizabeth Oak, 84–85
Queen's Green Canopy, 208

R

Rackham, Oliver, 84, 91, 100, 162
rainforest, Celtic, 117–19
Rainforest Alliance, 173
Ramsey, Bella, 199
Redfern, Margaret
  British Plant Galls (with Shirley), 46
  A Guide to Plant Galls in Britain (with Shirley), 46
red oak, 8
red squirrel (*Sciurus vulgaris*), 178, 192
redstart, 22
Reed, Tom, 180
regeneration, natural, 202–4
Remedy Oak, 141
reproduction, 187–88
Resilient Young Minds, 143–44
rhododendron, common (*Rhododendron ponticum*), 167
robin, 27, 28, 47, 173
Robin Hood, 31–32, 99
roots, 56–57, 150–52, 180

Roy, Arundhati, 107, 209
The Royal Oak (The Boscobel Oak), 78–79, 80
Royal Oak Day (Oak Apple Day), 80–81
Royal Oak pubs, 79–80
Royal Society for the Protection of Birds (RSPB), 15, 19, 206, 214, 216
royalty, and oaks
  about, 76–78
  The Bruce Tree, 83
  Charles II and The Boscobel Oak, 78–79
  Charles III and, 77, 80, 208
  Edward VI and the Remedy Oak, 141
  Elizabeth I and, 77, 83–86, 193
  Elizabeth II and, 77, 84, 208
  John and The Parliament Oak, 33, 77
  Oak Apple Day and, 80–81
  The Owen Glendower Oak, 33, 82–83
  Royal Oak pubs, 79–80

S

saproxylic beetles, 91–92, 102, 103. *See also* beetles
Savernake Forest (Wiltshire), 115–16
Schama, Simon, 194
Schneidau, Lisa, 120
sessile oak (Irish oak, Welsh oak, Cornish oak, durmast oak, *Quercus patraea*), 7, 9–10, 33
sex, oak, 187–88
shaggy strap lichen (*Ramalina farinacea*), 125, 131
Shakespeare, William, 6, 40, 48
  As You Like It, 51
  Macbeth, 54, 55, 100, 190
  The Merry Wives of Windsor, 51–52

Shardlow, Matt, 73
Shaw, George Bernard, 185
Sherwood Forest
  ancient oaks, 34–35
  The Bee Tree, 31, 34
  Budby Heath, 92
  generation gap in, 105
  historic extent, 100
  The Major Oak, 31–32, 98, 102, 180
  The Man-eating Caterpillar Tree, 34–35
  The Medusa Oak, 35, 88, 89–90, 91
  Moaning Mabel, 34
  Nottingham, linking with, 101
  The Parliament Oak, 33, 77, 103, 149
  Stumpy, 34
  in summer, 88
  woodpeckers, 14–15, 19
shipbuilding, 193–94
Shirley, Peter
  British Plant Galls (with Redfern), 46
  A Guide to Plant Galls in Britain (with Redfern), 46
silk button galls, 45–46, 88
silk button gall wasp (*Neuroterus numismalis*), 45–46
Simard, Suzanne, 150
Skinner Trap, 69
slugs, 129–30
Smith-Wright, Jim, 148
smoking, with wood, 138
snails, 127
Snow, Dan, 144
soil, 155–57, 157–58, 179–80
Somerset, 37–38
song thrush, 27–28, 28–29
spangle galls, 44–45
spangle gall wasp (*Neuroterus quercusbaccarum*), 45
Speight, Beccy, 190, 213

spiders, 41, 72, 127
spirituality
    devil, summoning, 115–16
    Druids, 110–11
    fairies and other spirits, 110, 112–13, 120
    gods, associated with oaks, 111–12
    gospel oaks, 113–14
    magic, 114–15, 126
    Saint Boniface and the oak tree, 113
spotted longhorn beetle (black-and-yellow longhorn, *Rutpela maculata*), 92–94, 95
spring, 224–25
springtails, 157–58
squirrels, 177–78, 191–92, 226
Stanley, Kevin, 208
starling, 22–23
The Strangling Tree, 116
The Strathlevan House Oak (The Bruce Tree), 83
Strutt, Jacob George, 83
Stump Up For Trees, 21
Stumpy (ancient oak), 34
suffragettes, 32
summer, 225–26
summer bolete (*Boletus reticulatus*, *Boletus aestivalis*), 145, 149–50
summer truffle (*Tuber aestivum*), 147
symbiosis, 152, 153
Synergus gallaepomiformis, 40

T

Ta-bu-ce, 135–36
tannin, 138, 142, 148
tawny owl, 226
Taylor, Andy, 150–52, 153–54, 156, 157
Tea Party Oak, 33
Tennyson, Alfred, Lord
    'The Talking Oak', 160

threats, to oaks
    acute oak decline (AOD), 161, 163–65
    chopping up landscapes, 172–73
    climate change, 155, 157, 164, 173–74, 196–97, 204
    deer, 176–77
    development, 169–71
    invasive (non-native) plants, 167–68
    oak jewel beetle, 161, 163–65
    oak processionary moth, 165–67
    pollution, 71, 124–25, 128, 174–75
    soil compaction, 179–80
    squirrels, 177–78
    See also conservation
Thunberg, Greta, 181, 183
time, and trees, 57
Tolkien, J. R. R.
    The Lord of the Rings, 55
Torymus nitens, 41
treading lightly, 179–80, 215
Tree, Isabella, 99–100, 203, 209
Tree Bingo (game), 184
Tree Charter (Charter for Trees, Woods and People), 46–47
The Tree House Oak, 43
tree lungwort (*Lobaria pulmonaria*), 123
trees
    ailing trees, tips, 217–18
    buying, tips, 214
    conservation promises, 197–98
    diseases and pests, 166–67, 214
    generation gap, 104–5
    meeting a tree, tips, 220–23
    planting vs. natural regeneration, 202–4
    time and, 57
    veteranisation, 105–6
    See also forests and woods; oaks

Trioza remota galls, 42–43
Trotter, Torrens, 84
Turkey oak (*Quercus cerris*), 8, 39, 45
The Turner Oak, 179, 218
Tweet of the Day (BBC Radio 4), 27, 29

U

umbellifers, 93, 95
United Kingdom, 181, 196–97
usnic acid, 123

V

veteranisation, 105–6
veteran trees, 12, 104
Virgil
  Georgics, 56, 57
volunteers, 206, 208, 214–15

W

wasps
  Synergus gallaepomiformis, 40
  Torymus nitens, 41
  See also gall wasps
weasels, 25
well-dressing, 124
Welsh oak. See sessile oak
Wentwood Forest, 208
West, Samuel, 26
whisky, 139–40
White, Elyse, 199, 202, 207
white oak (*Quercus alba*), 137
Whittle, Lorienne, 187
The Wildlife Trusts, 101, 172–73, 206, 214, 225
Windsor Forest, 52–53
Windsor Great Park, 53, 77, 85, 193
winter, 223–24
winter moth (*Operophtera brumata*), 23, 67
Winterson, Jeanette, 48
witchcraft and magic, 114–15, 126
Wohlleben, Peter, 132, 150, 154
wolves, 176, 177
wood
  dead wood, 89–91, 91–92, 96–97
  heart wood, 90–91, 147–48
  for smoking, 138
The Woodland Trust
  about, 206, 214
  free tree packs, 217
  Hainault Forest and, 195
  on HS2 rail development, 170
  on legal protections for ancient trees, 205
  Nature's Calendar, 173, 213, 225
  on non-native plants, 168
  oak leaf logo, 58–59
  on pollution, 174–75
  Resilient Young Minds and, 143
  'State of the UK's Woods and Trees' report, 97, 181
  Treescapes, 101, 173, 203
  Young People's Forest and, 65
woodpeckers, 14–18, 19–20, 24, 25, 30, 47, 106, 224
Woods, Ray, 156
woods and forests, 97, 98–102, 143–45, 156. See also trees
worms, 218
wren, 29
Wright, John, 146
wryneck (*Jynx torquilla*), 18–19
Wyatt, John, 138
Wyver, Jamie, 189

X

xeno-canto (website), 29
Xylella fastidiosa, 161

Y

Yahr, Rebecca, 122
Yardley Chase (Northamptonshire), 57

Young, Barbara, 197
Young, Francis Brett, 13
Young People's Forest
    (Derbyshire), 65, 198–200,
    200–202, 207

Z

Zeus, 111–12